3rd Edition

JEFF **GREEN** / ROGER **GREEN**

AUSTRALIA AND NEW ZEALAND

AUTOMOTIVE AIR CONDITIONING

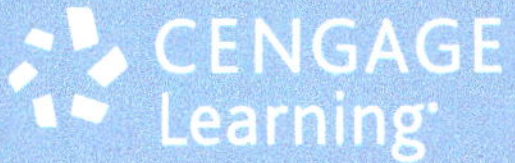

Australia • Brazil • Japan • Korea • Mexico • Singapore • Spain • United Kingdom • United States

Automotive air conditioning: Australia and New Zealand
3rd Edition
Jeff Green & Roger Green

Publishing manager: Dorothy Chiu
Publishing editor: Fiona Hammond
Publishing assistant: Miriam Allen
Developmental editor: James Forsyth
Project editor: Tanya Simmons
Text designer: Sarah Anderson
Cover design: Danielle Maccarone
Art direction: Danielle Maccarone
Permissions/Photo researcher: Wendy Duncan
Editor: Edward Caruso
Proofreader: Kevin Young
Indexer: Julie King
Cover: Shutterstock.com
MPS Limited

For product information and technology assistance,
in Australia call 1300 790 853 ;
in New Zealand call 0800 449 725

For permission to use material from this text or product, please email **aust.permissions@cengage.com**

National Library of Australia Cataloguing-in-Publication Data
Green, Jeff, 1951-
Automotive air conditioning : Australia and New Zealand/Jeff Green and Roger Green.
3rd ed.
9780170197748 (pbk.)
Automobiles--Australia--Air conditioning.
Automobiles--New Zealand--Air conditioning.
Automobiles--Australia--Air conditioning--Maintenance and repair.
Automobiles--New Zealand--Air conditioning--Maintenance and repair.
Green, Roger, 1954-
629.2772

Cengage Learning Australia
Level 7, 80 Dorcas Street
South Melbourne, Victoria Australia 3205

Cengage Learning New Zealand
Unit 4B Rosedale Office Park
331 Rosedale Road, Albany, North Shore 0632, NZ

For learning solutions, visit **cengage.com.au**

Printed in Australia by Ligare Pty Limited.
4 5 6 7 8 9 24 23 22 21

BRIEF CONTENTS

Section 1
Air-conditioning theory

Section 2
System diagnosis

Section 3
Service procedures

CONTENTS

2.2 System-diagnosis charts

2.3 System-diagnosis problem sheets

2.4 Pressure/temperature charts

2.5 Temperature–pressure exercises

2.6 System-diagnosis exercises

Section 3
Service procedures

3.1 Service procedure 1 Connecting the manifold and gauge set to the air-conditioning system 186

ACKNOWLEDGEMENTS

As in previous editions, this book has been developed from learning and research as part of our dedication to lesson preparation for teaching our air-conditioning classes. Each chapter and exercise in this book has been tested and developed over many years to enhance the ease of learning, and consolidation of previously gained skills.

We thank all the professionals and colleagues involved for their contributions, especially the people that allowed us to incorporate the diagrams and pictures used in this and previous publications, and all those who placed their information on the Internet for the great contribution this made to our research. We would also like to thank our students. At times they were the guinea pigs in our implementation of these exercises. Through them we were able to develop the easy-to-use format of this material.

Cengage Learning would like to thank the following people for their reviewing and contributions:

› John O'Grady, Director, Contract Environmental Ltd, Auckland
› Mike Dubey, Senior Lecturer in Air Conditioning at the Manukau Institute of Technology in Auckland
› Shane Williams, South Australian State Manager, Careers Australia
› John Henry, formerly Leading Vocational Teacher and Business Manager, SkillsTech Australia
› Bruce Cock, Independent Automotive Air Conditioning Consultant
› Shaun Johnson, Automotive Instructor, Central Gippsland Institute of TAFE (GippsTAFE)
› Noel Woodbury, Teacher of Light Automotive, New England Institute, TAFE NSW
› Robert Clark, Lecturer (Refrigeration & Air Conditioning), Polytechnic West College

ABOUT THE AUTHORS

Jeff Green

Jeff recently retired after 27 years of teaching with TAFE NSW as an automotive teacher, and after 42 years of industry experience. He has specialised for over 23 years in the teaching and training of automotive air-conditioning courses. During this time he has written and proofed several air-conditioning courses for the NSW TAFE system and constructed the practical and theory examinations for these courses' accreditation. He also developed short courses for the recognition of prior learning, work-skills and industry experience for the Certificate II qualification in air-conditioning. This allows technicians to obtain an air-conditioning licence through the Australian Refrigeration Council (ARC). Jeff also drafted the first code of practice for the Vehicle Air-conditioning Specialists of Australia (VASA).

During his career Jeff has kept abreast of technical changes and industry requirements, and developed lessons to incorporate the latest industry advances as they were implemented. In this latest book, Jeff and co-author Roger have sourced and incorporated the latest industry developments.

Roger Green

Roger Green is currently Head Teacher of Automotive and Auto Electrical at St George TAFE College at Kogarah, a position he has held for 12 years. Roger has over 22 years experience as a teacher in TAFE NSW and has specialised in Automotive Air Conditioning for 18 years. Roger stays up to date with industry developments and is at the forefront of recent changes in design and new refrigerants in the air-conditioning industry.

He has a Diploma in Teaching (Tech.), a Graduate Diploma in computer-based learning, and is a member of both the Institute of Automotive of Mechanical Engineers (IAME) and Society of Automotive Engineers Australasia (SAE-A).

RESOURCE GUIDE

For the student

As you read the text, you will find a wealth of features designed to assist your understanding of the theory of automotive air conditioning and its real-life application.

Section 1: Air-conditioning theory

Section 1 covers the knowledge base that you'll need. Every chapter in Section 1 has a list of **Objectives** that suggest the key learning outcomes covered throughout the chapter. Each chapter also has a list of **essential knowledge and associated skills** to support your progress through your training competencies.

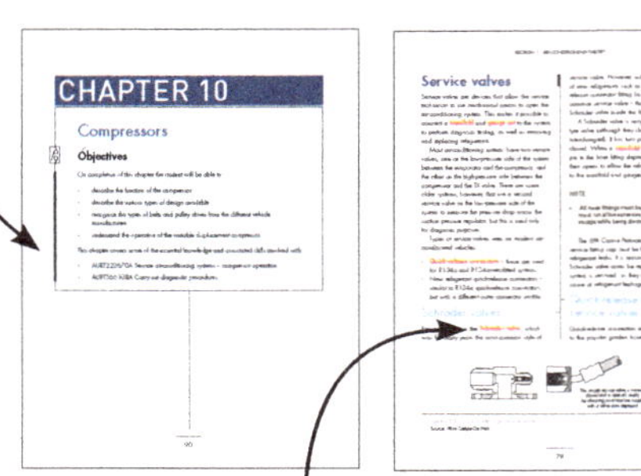

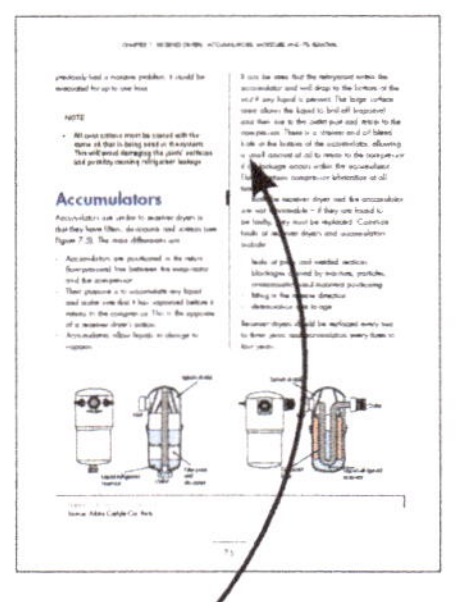

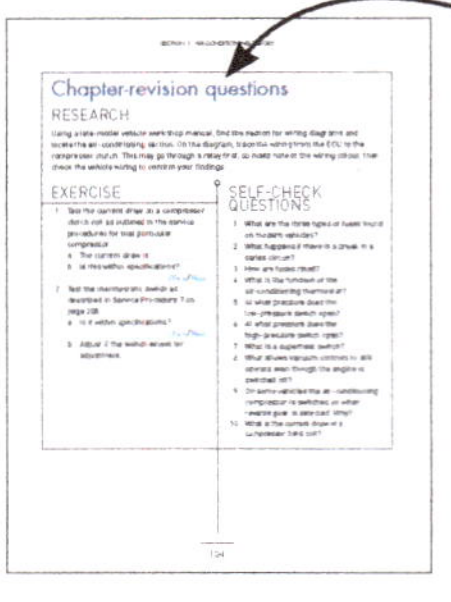

Each end-of-chapter feature in Section 1 has a series of **self-check questions**, which enable you to quickly test your own understanding of the chapter you have just read – **answers** are at the back of the book. Where relevant, the theory chapters are also linked to the most appropriate applications through the **exercise** sections and there are tips for practical ideas for workshop **research**.

Key terms and important definitions are bolded in the text and appear in the **glossary** at the end of the book for your quick reference.

Instructions and procedures that require special emphasis, such as for safety reasons, are written in **notes** boxes.

Section 2: System diagnosis

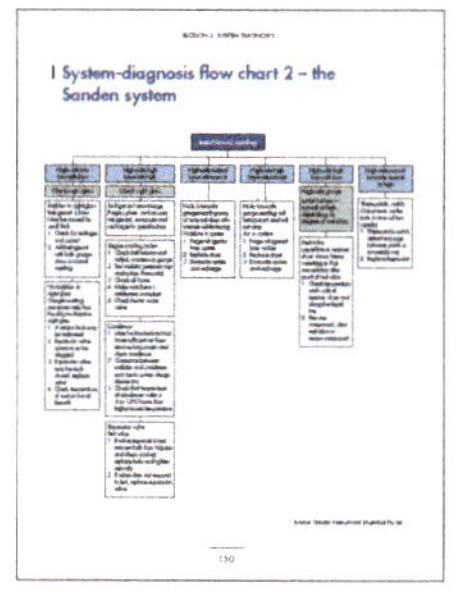

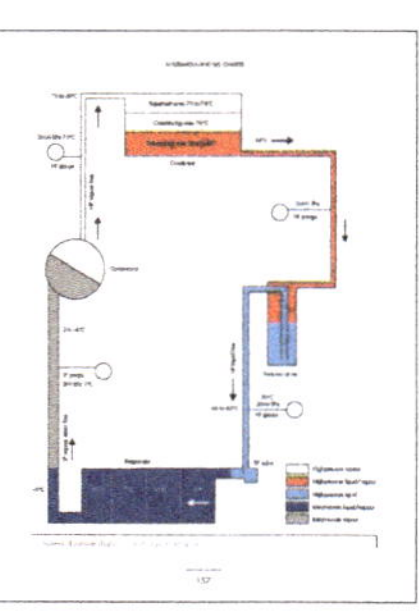

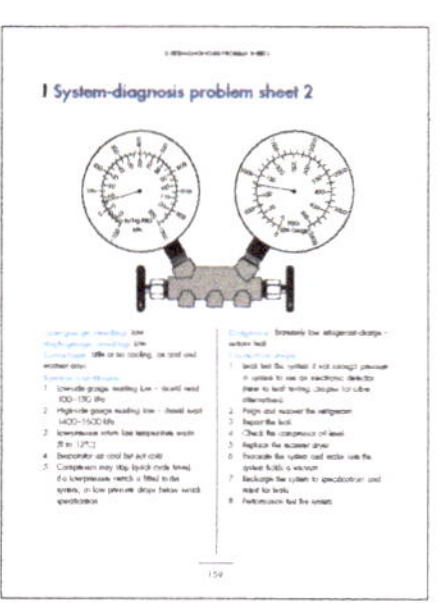

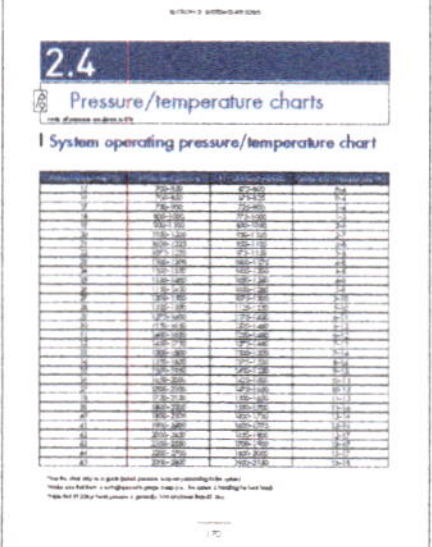

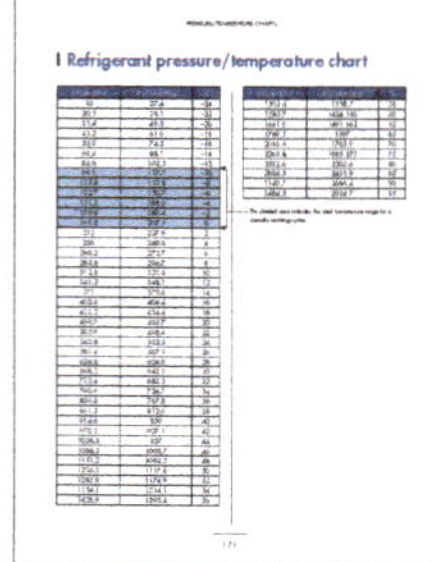

Section 2 includes important reference charts and tables for your own diagnosis of the air-conditioning system. These charts include **system-diagnosis flow charts**, common **system-diagnosis problem sheets**, **system operating pressure/temperature charts** and **refrigerant pressure/temperature charts**.

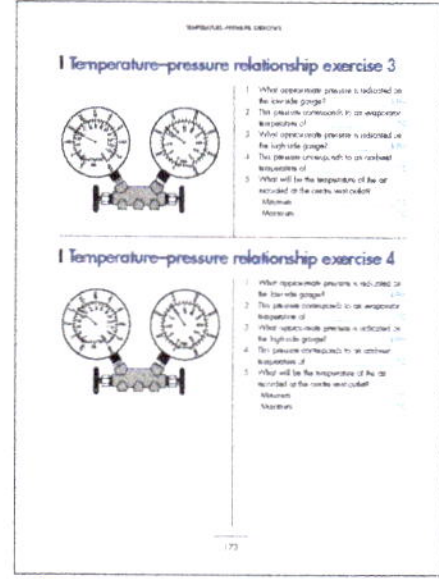

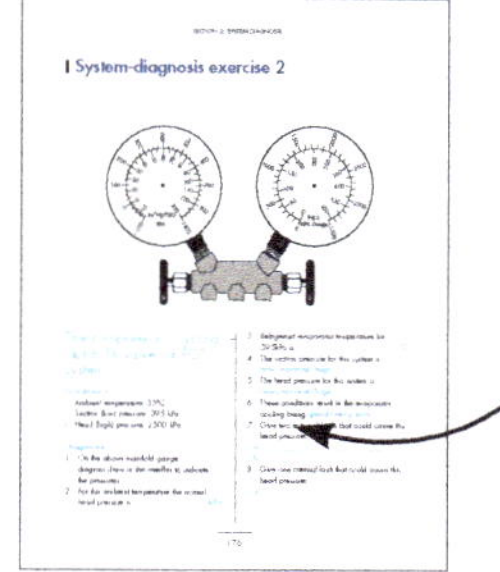

In Section 2, you will also find exercises to help you apply your understanding of the diagnosis charts to practical workshop situations, helping you to work through **temperature–pressure exercises** as well as **system-diagnosis exercises**.

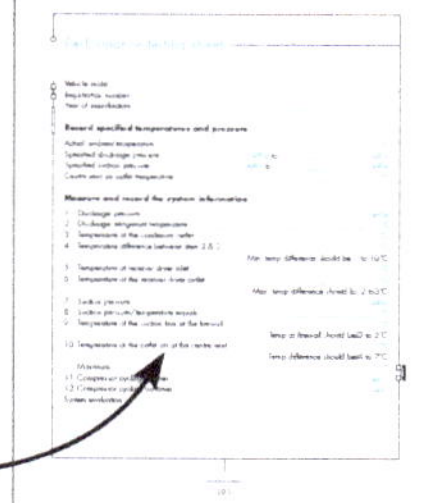

Section 3: Service Procedures

Section 3 provides a guide to the most common and useful servicing and repair procedures with easy to follow, step-by-step instructions. There is a convenient **performance testing sheet** included.

Student online resources

Log in to **http://www.cengagebrain.com** for access to the *Automotive Air Conditioning* student companion website, which features web links to current industry and training websites, templates of the worksheets and performance-testing sheets, as well as online activities.

For the instructor

Cengage Learning is pleased to provide you with a selection of electronic and online resources to help you teach in automotive air conditioning. These teaching tools are available on the Instructor website accessible by visiting **http://login.cengage.com**.

Instructor's manual

The Instructor's manual provides you with content to help set up and administer your automotive air-conditioning subject. It includes a list of the chapter specific learning objectives, the key terms within each chapter and solutions to end-of-chapter questions. The Instructor's manual includes completed versions of the **pressure/temperature exercises** and the **system-diagnosis exercises**.

Mapping grid

The mapping grid that provides an overview of the key units of competency covered in *Automotive Air Conditioning* is available for your download.

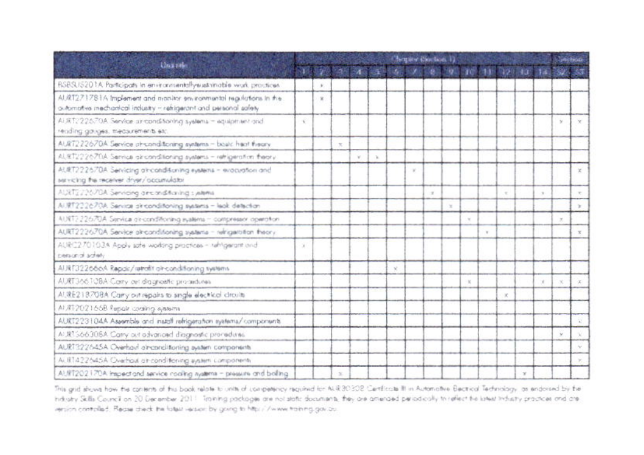

Unit title	Chapter (Section 1)														Section	
	1	2	3	4	5	6	7	8	9	10	11	12	13	14	S2	S3
BSBSUS201A Participate in environmentally sustainable work practices		x														
AURT271781A Implement and monitor environmental regulations in the automotive mechanical industry – refrigerant and personal safety		x														
AURT222670A Service air-conditioning systems – equipment and reading gauges, measurements etc	x														x	x
AURT222670A Service air-conditioning systems – basic heat theory			x													
AURT222670A Service air-conditioning systems – refrigeration theory				x	x											
AURT222670A Servicing air-conditioning systems – evacuation and servicing the receiver dryer/accumulator							x									x
AURT222670A Servicing air-conditioning systems								x				x		x		x
AURT222670A Service air-conditioning systems – leak detection									x							x
AURT222670A Service air-conditioning systems – compressor operation										x					x	
AURT222670A Service air-conditioning systems – refrigeration theory											x					x
AURC270103A Apply safe working practices – refrigerant and personal safety	x															
AURT322666A Repair/retrofit air-conditioning systems						x										
AURT366108A Carry out diagnostic procedures										x				x	x	x
AURE218708A Carry out repairs to single electrical circuits												x				
AURT202166B Repair cooling systems																
AURT223104A Assemble and install refrigeration systems/components																x
AURT366308A Carry out advanced diagnostic procedures															x	x
AURT322645A Overhaul air-conditioning system components																x
AURT422645A Overhaul air-conditioning system components																x
AURT202170A Inspect and service cooling systems – pressure and boiling			x										x			

This grid shows how the contents of this book relate to units of competency required for AUR30308 Certificate III in Automotive Electrical Technology, as endorsed by the Industry Skills Council on 20 December 2011. Training packages are not static documents, they are amended periodically to reflect the latest industry practices and are version controlled. Please check the latest version by going to http://www.training.gov.au

Artwork files

These digital files of pictures, tables and graphs from the text can be used in a variety of media. Use them in WebCT or Blackboard, PowerPoint presentations or copy them onto overheads.

Section 1
Air-conditioning theory

CHAPTER 1

Introduction to air conditioning, metric conversions and refrigerant handling regulations

Objectives

On completion of this chapter the student will be able to:

- relate the history of automotive air conditioning
- define the term 'air conditioning'
- use given safety precautions when working with refrigerants
- use the formulas provided to convert imperial units to metric units and metric units to imperial units
- understand the licensing and regulations required to work on air-conditioning systems
- apply for the appropriate licences required.

This chapter covers some of the essential knowledge and associated skills involved with:

- AURC270103A Apply safe working practices – refrigerant and personal safety
- AURT222670A Service air-conditioning systems – equipment and reading gauges, measurements etc.

Introduction

The purpose of this publication is to allow the automotive air-conditioning trainee or student to gain a better-than-basic working knowledge of the system and its components plus a diagnostic ability. The student should be able to apply this knowledge so as to competently repair and/or modify an air-conditioning system in order to return it to its full working capacity.

Refrigeration and air conditioning are not discoveries of the past century. Simple forms of refrigeration and air conditioning have been in use for thousands of years.

Within the past 70 years many aspects of modern life have been made possible through the development of sophisticated air-conditioning systems. Many components associated with the worldwide space program could not have been possible without the use of air conditioning. These components were manufactured in a temperature and **humidity**-controlled environment. Also, modern medicine and delicate machine components are perfected in scientifically controlled air-conditioned environments.

Computer centres are only able to function properly because they are kept within a specific temperature and humidity range, which depends on complex air-conditioning systems.

Automotive air conditioning became available in 1940, but did not start to become popular in Australia until the late 1960s. At this time the air-conditioning system was an add-on under-the-dash unit, of which the most popular was the Mark 4 system (see Figure 1.1).

Since then the number of vehicles fitted with air conditioning has greatly increased. During the early 1970s, **OEM** (original equipment manufacturer)-fitted integrated air-conditioning systems were introduced by each of the big three Australian car manufacturers.

Air conditioning is now the most popular accessory fitted to new vehicles. Some 95 per cent of new vehicles are now fitted with air conditioning.

What is air conditioning?

Automotive **air conditioning** is the process by which the air is cooled and cleaned, the humidity lowered and the air circulated. The quantity and quality of the air is also controlled. Under ideal conditions the air-conditioning system can be expected to accomplish all these tasks at the same time. The air-conditioning system in modern vehicles is designed to lower the temperature to

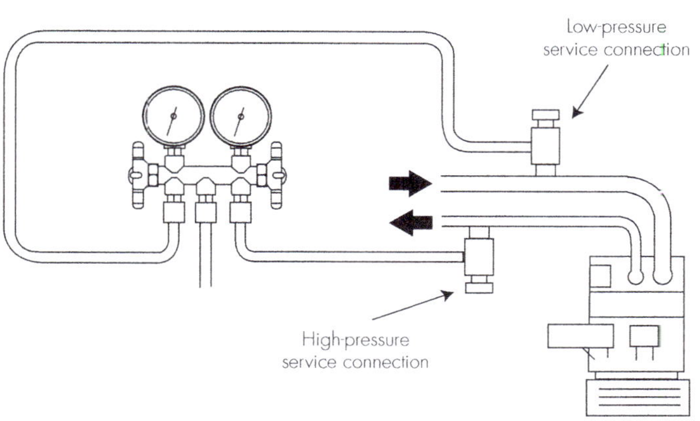

Figure 1.1 Early under-the-dash mounted unit
Source: Smiths Industries Pty Ltd Australia

22–25°C and the humidity to approximately 50 per cent.

The air-conditioning student must have a thorough understanding of:

- each system air-conditioning component
- the extra demands placed on the cooling system
- **refrigerant** flow and change of state
- system electrical circuits
- system controls.

Knowledge of the available equipment, special tools, and diagnosis and repair techniques is also essential. These will all be discussed in detail in their own sections.

History and development of the automotive air-conditioning system

The first automotive air-conditioning unit made its debut in 1927. At that time 'air conditioning' meant that a vehicle was fitted with a heater and a filtered air ventilation system.

In 1940 Packard offered the first method of cooling a car by means of refrigeration. In reality, this involved a commercially developed air-conditioning unit adapted for automotive use. In the 1950s demand for air-conditioned vehicles began to rise, especially in the warmer states of America.

The popularity of air conditioning for vehicles grew quickly in Australia after it had been introduced as an after-market unit in the late 1960s. Today, mobile air conditioning is available for every form of motor vehicle and transport equipment found and operated within both countries. From this brief history it can be seen that there has been rapid growth in the mobile air-conditioning industry over the past 70 years, and especially over the past 40 to 50 years.

Cost of operation

An air-conditioning system places extra load on the engine and its cooling system – and any extra engine load requires extra fuel. Therefore, an air conditioner should increase costs. It does marginally. This mostly occurs in city stop-start driving conditions. Developments in engine management systems have improved engine economy and power output and so reduced or even offset, the losses of a vehicle using an air-conditioning system.

At highway speeds, with the windows closed (and a consequent reduction in wind drag) and the air-conditioning system operating normally, there is actually 2–3 per cent efficiency saving on average over a non-air-conditioned vehicle with the windows open.

With the continuing popularity of OEM air-conditioned vehicles, and as we keep our vehicles for an average of eight to ten years, it is obvious that the need for the installation of air-conditioning units has decreased, but their maintenance needs have greatly increased.

Safety precautions when handling refrigerants

Remember that all refrigerants are dangerous. They are also:

- odourless
- colourless
- heavier than air
- able to freeze on contact with skin and eyes
- oxygen depleting.

It is not possible to cover every safety precaution that an air-conditioning technician needs to be aware of with each of the different types of refrigerants found in the automotive industry at the present time. However, general refrigerant safety precautions are described below.

- All refrigerants must be properly stored.
- All refrigerants released under pressure will have a temperature of between −30°C and −26°C. Therefore, frostbite will occur if escaping refrigerant contacts the skin. Blindness may result from contact with the eyes.

The following procedures must be adopted when working with refrigerants:

- Eye protection (safety glasses) should be worn.
- Overalls or suitable industrial clothing with long sleeve shirts must be worn.
- Suitable hand protection is advisable.

If refrigerant does strike the eyes or skin, the following first-aid treatment must be immediately administered:

- Splash large quantities of cool tap water on the area for several minutes.
- Cover the area.

You should then seek medical advice.

NOTE

+ If refrigerant is splashed in the eyes, the injured worker must be treated *immediately*, as blindness can occur.
+ All refrigerants are heavier than air and can cause asphyxiation. Keep the work area well ventilated.
+ A poisonous gas is created when refrigerants contact a naked flame or very hot metals. Breathing this gas can cause severe illness.
+ Many refrigerants ignite when they come into contact with a flame as flammable refrigerants are becoming commonplace.
+ With the number of refrigerants on the market and the variety of chemicals used to produce them, it is essential to consider all refrigerants as potentially flammable and toxic when they come into contact with hot metals or naked flame.
+ Refrigerant containers must be stored in a cool area. Heating the container or placing it in direct sunlight will increase the internal temperature and, therefore, cause its pressure to rise, which can be dangerous. Temperatures above 54°C must be prevented.
+ Refrigerant containers are not always colour-coded to identify the refrigerant they contain. Therefore, the containers must be clearly labelled, and only the refrigerant named on the label is to be put in that container.

The metric system

Australia converted to the S.I. (Système International) or, as it is more commonly known, the metric system, in the early 1970s. As some of the vehicles driven on Australian roads, and some equipment may still use the imperial system (e.g. vehicles and equipment imported from the US), knowing how to convert from imperial to metric (and vice versa) is vital.

The air-conditioning technician only needs to know a few conversions, such as weight, pressure, heat and **vacuum**.

Weight

Refrigerant is always measured by weight. The basic metric unit of weight is the gram (g).

1000 grams = 1 kilogram (kg).

The imperial unit of weight is the **ounce** (cz).

16 ounces = 1 **pound** (lb).

The relevant conversion factors are:

ounces to grams: x ounces by 28.35
grams to ounces: ÷ grams by 28.35
pounds to kilograms: x pounds by 0.45359

Table 1.1 gives an outline for converting weight.

Table 1.1 Weight conversion table

Grams	Ounces	Pounds
50	1.76	0.11
100	3.52	0.22
200	7.05	0.44
400	14.11	0.88
500	17.64	1.20
800	28.22	1.76
1000	35.27	2.2
1500	52.91	3.3
2000	70.55	4.4

System oil capacities

The relevant conversion factors are:

28.35 cc, grams or ml = 1 oz, so therefore
150 cc, grams or ml = 5.29 oz
175 cc, grams or ml = 6.17 oz

Pressure

One of the most important measurements that an air-conditioning technician will make involves the pressures within a working air-conditioning system. These pressures help to determine the operating conditions within the system and therefore diagnostic procedures can be performed. Pressure and temperature are related within the air-conditioning system. The metric unit of pressure is the pascal (Pa).

1000 pascals = 1 kilopascal (**kPa**]
100 kilopascals = 1 **bar**

Although these metric pressure units are the most common, some manufacturers use kilograms/square centimetre (**kg/cm²**), bar or the non-metric **psi** (pound per square inch).

Table 1.2 is a pressure-conversion table air-conditioning students will need to study.

Table 1.2 Pressure conversion table

1 kPa kPa	÷ 100 (=bar)	÷ 98 (=kg/cm²) bar	÷ 6.895 (=psi) kg/cm²	psi
100	1	1.02	15	14.5
200	2	2.04	29	29
300	3	3.06	44	43.5
400	4	4.08	58	58
500	5	5.10	73	72.5
600	6	6.12	87	87
700	7	7.14	102	101.5
800	8	8.16	116	116
900	9	9.18	131	130.5
1000	10	10.20	145	145

Heat

There are several aspects of heat that should interest the air-conditioning student:

› *Quantity of heat* (or more accurately 'energy transferred') is measured in *calories*.
  » 1 calorie is the amount of heat required to raise the temperature of 1 gram of water by 1°C
  » The joule (J) is also used as a measurement of heat energy.
  1 calorie = 4.180 **joules**

Note that it is impossible to measure the quantity of heat without special equipment in a laboratory.

- *Intensity of heat* (also known as **sensible heat**) is measured in degrees **Celsius** (°C). The Celsius scale is based on the fact that water freezes at 0°C and boils at 100°C.

Another common measure of heat intensity uses degrees **Fahrenheit** (°F) as its basic unit. To convert Fahrenheit to Celsius, use the following formula:

°C = 0.555 x (°F – 32)

To convert Celsius to Fahrenheit uses this formula:

°F = (1.8 x °C) + 32

When working with air-conditioning systems it is important to know the temperature of the day (**ambient air temperature**) as this determines the operating pressure found in the system. **Latent heat** is important to the air-conditioning technician as it refers to the large amount of heat transfer as the refrigerant condenses or evaporates. Latent is derived from the Greek. It means 'hidden', so we cannot see it, and is defined as a **change of state** without a change in temperature. For example, a block of ice as it melts does not increase in temperature while it is melting.

Note that the sum of both latent heat and sensible heat is known as enthalpy, the units being joules.

Vacuum

The metric system commonly measures **vacuum** as minus kPa (–kPa), but it can also be measured in millimetres of mercury (**mm/Hg**) or micrometers of mercury (mμ/Hg).

The imperial measurement for vacuum uses inches of mercury (**in/Hg**). To convert inches of mercury to kPa:

in/Hg x –3.386 = –kPa

Note that the answer is in minus kPa.

To convert kPa to inches Hg:

–kPa ÷ –3.386 = in/Hg

Note that the answer is positive, as two negatives make a positive.

Microns are also used to measure the vacuum in some test equipment. At **atmospheric pressure** the micron scale will show 760 000 microns and at –100kPa it will read 200 or less microns, depending how dry the system is.

Table 1.3 is another that trainees will need to study.

Table 1.3 Vacuum conversion table

–kPa	in/Hg	mm/Hg
-10	3	685
-20	6	612
-30	8.9	530
-40	11.8	455
-50	14.8	380
-60	17.7	310
-70	20.7	232
-80	23.6	158
-90	26.6	84
-100	29.5	10
-101.6	30	0

Government legislation

In Australia there is a government agency called the Australian Refrigeration Council, which operates under the Department of Sustainability, Environment, Water, Population and Communities. They have introduced legislation requiring air-conditioning technicians to have the appropriate equipment and service records to comply under the requirements of the law.

In future, the New Zealand government may also set up appropriate legislation to meet these requirements. At present, all refrigeration technicians must meet the Hazardous Substances and New Organisms (HSNO) Act requirements.

The Australian regulations require all technicians to:

- be trained and competent (Aust. code of practice Part A.2)
- have the appropriate equipment (Aust. code of practice Part A.5)
- maintain the equipment to manufacturers' requirements (Aust. code of practice Part A.5.2 to A.5.4 and A.7)
- capture all refrigerant for reuse or destruction (Aust. code of practice Part A.6) – in New Zealand this is a requirement of the *Ozone Depletion Act 1996*
- keep the appropriate paperwork to show refrigerant bought and captured and equipment serviced (Aust. code of practice Part A.5.3, A.9.5).

The New Zealand HSNO Act and Regulations requires all parties using and/or storing certain hazardous substances (including flammable substances) above certain quantities to become approved handlers. Locations where these substances are stored or used may also need to hold Location Test Certificates.

Further details of the legislation are available:

- for Australia: http://www.arctick.org/index.php (select 'Code of Practice' then 'Automotive code of Practice')
- for New Zealand: http://www.mfe.govt.nz/issues/ozone/depleting-substances/cfc-mgmt-strategy.html

There are severe penalties for handling refrigerant without the relevant licences; businesses can be fined up to $5500 and individuals up to $1100. The Arctick website has a number of fact sheets that can be printed and kept for reference. Although at the time of writing New Zealand does not have a licensing scheme they still operate under a Code of Practice.

In Australia there is a requirement that anyone who works on or handles refrigerant must be licensed. The appropriate licences for automotive technicians are as follows:

- licence CL100 – this is for students who have no licensed tradesman at work to supervise them and is for classroom training only.
- licence TL000 – this is for trainees who are able to use their licence at work under supervision and in the classroom while undertaking training.

NOTE

- Evidence of either of the above training licenses must be produced when training is undertaken.
- When applying for the AAC02 Automotive Air Conditioning licence, applicants must produce evidence of having completed Certificate 11 in automotive air conditioning.
- The AAC02 Automotive Air Conditioning licence will be restricted if the subject AURT422645A Overhaul air conditioning has not been completed during training.

IMPORTANT NOTE

- Before continuing with this book or undertaking any training within Australia, students are required go to online to http://www.arctick.org. Please read and download the 'Australian automotive code of practice', and then fill out the appropriate licence form. Different parts of the code of practice will be used where appropriate within the various chapters of this book.

Chapter-revision questions

SELF-CHECK QUESTIONS

1. When was air conditioning first available in vehicles?
2. What four areas must an air-conditioning system control?
3. What are some compulsory safety precautions that need to be observed when working on air-conditioning systems?
4. If refrigerant contacts the skin or eyes what is the initial first aid treatment?
5. Are refrigerants lighter or heavier than air?
6. What are two precautions to be observed concerning refrigerant containers?
7. What does S.I. refer to?
8. In Australia and New Zealand how is refrigerant quantity measured?
9. If an American vehicle had an air-conditioning system with an oil capacity of 6 fl. ounces what would this be in grams?
10. What two areas of heat need to be considered when working on air-conditioning systems?
11. How is vacuum measured in the metric system?
12. What are the main requirements under government legislation that air-conditioning technicians are required to comply with?
13. What is the maximum penalty for an individual who handles refrigerant without a licence under current legislation?
14. What is the maximum penalty for a business that handles refrigerant without a licence under current legislation?
15. Is a student technician required to have a licence when training in a classroom?

CHAPTER 2

Climate change

Objectives

On completion of this chapter the student will be able to:

- state the role of the ozone layer
- describe the composition of the atmosphere
- describe how the ozone layer is being damaged by chlorine in refrigerants
- define the term 'ozone hole'
- describe the difference between 'greenhouse effect' and 'ozone depletion'
- state the effects that the ozone depletion is having on the environment
- understand the concept of sustainability.

This chapter covers some of the essential knowledge and associated skills involved with:

- AURT271781A – Implement and monitor environmental regulations in the automotive mechanical Industry – refrigerant and personal safety
- BSBSUS201A – Participate in environmentally-sustainable work practices.

Climate change and ozone depletion

There are many factors that contribute to both climate change and **ozone** depletion. These are different but related. Some gases, such as carbon dioxide, methane, nitrous oxide and hydroflurocarbons (HFCs) contribute to climate change; some gases contribute to ozone depletion, such as chlorofluorocarbons (**CFCs**), hydrochlorofluorocarbons (HCFCs) and halons; and some gases contribute to both (e.g. CFCs, HCFCs).

Since the discovery of the depletion of the ozone layer, as evidenced in the ozone layer over Antarctica, there has been widespread concern about its effect on human health and the environment. Ozone depletion, together with climate change (often called global warming), has attracted widespread media attention and well-founded global concern.

This chapter will explain the importance of the ozone layer, how it is being affected and depleted, what is depleting it, and what industry and governments are doing to correct the damage.

What is ozone?

Many years ago, it was believed that ozone was abundant at the seaside and healthy to breathe. It was thought that this contributed to the enjoyment of being at the beach. Many hotels and guest-houses at seaside resorts were named 'Ozone', and even the *Concise Oxford Dictionary* described ozone as having a pungent refreshing odour and, figuratively, an exhilarating influence. In fact, ozone in large concentrations in the earth's lower atmosphere is a poisonous gas, but in its proper place (in the **stratosphere**) it does protect life from damaging ultraviolet **radiation**.

Ozone does have a pungent odour, scientifically described as 'irritating'. In high concentrations it has a blue colour, in contrast to oxygen, which is colourless, tasteless and odourless. Ozone is created in the lower atmosphere as a result of pollution. It can be seen, together with oxides of nitrogen and **hydrocarbons**, as photochemical smog. It is a form of oxygen, each molecule containing three atoms of oxygen instead of the normal two atoms that are found in the oxygen molecules that we breathe.

The earth's atmosphere

The earth's atmosphere is a covering of gases that surround the globe. It is an enormous mass, equivalent to about one million tonnes for every person living on earth. The major gases in the atmosphere are nitrogen, an inert gas, which comprises 78 per cent by volume; and oxygen, which is vital for life and comprises 21 per cent. Water **vapour**, a portion of which is seen as clouds, accounts for less than I per cent (see Figure 2.1 and Table 2.1).

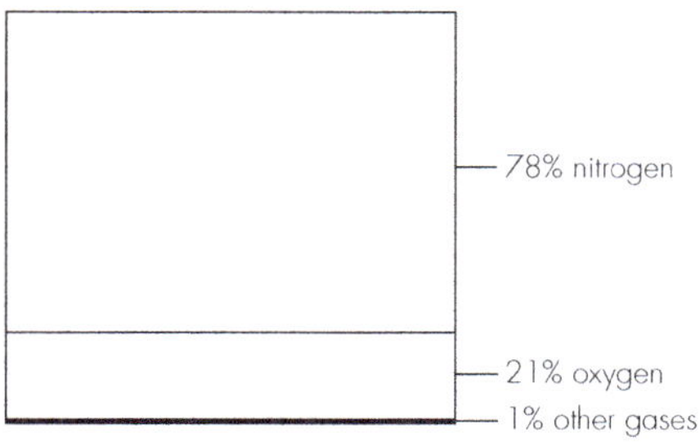

Figure 2.1 Component gases of the atmosphere

Table 2.1 Average composition of the atmosphere up to an altitude of 25 km

Gas Name	Chemical Formula	Percent Volume
Nitrogen	N_2	78.08%
Oxygen	O_2	20.95%
*Water	H_2O	0 to 4%
Argon	Ar	0.93%
*Carbon dioxide	CO_2	0.0360%
Neon	Ne	0.0018%
Helium	He	0.0005%
*Methane	CH_4	0.00017%
Hydrogen	H_2	0.00005%
*Nitrous oxide	N_2O	0.00003%
*Ozone	O_3	0.000004%

*variable gases

There are also several trace gases that, though small in volume, play critical roles in the atmosphere. Carbon dioxide is a trace gas that (as of 2011) has a concentration of only 390 parts per million by volume (0.0039 per cent ppmv), but which absorbs infrared radiation, thus warming the atmosphere through the phenomenon known as the **greenhouse effect**. Without its ability to retain this heat, the earth would be about 33°C colder and unable to support life. The increase of 25 per cent of this small amount of the total atmospheric composition of carbon dioxide over the past century is the primary cause of global warming of approximately 1–1.5°C to the year 2000.

Ozone, another trace gas, occurs at even lower concentrations in the lower atmosphere (about 0.4 ppmv) but it is essential for absorbing ultraviolet radiation from the sun, which is potentially very damaging to life on earth.

The atmosphere extends skywards for hundreds of kilometres. Its lowest part, up to about 15 km, is known as the **troposphere**. Above that, to an altitude of about 50 km, is the stratosphere. The troposphere is turbulent and contains clouds, winds, storms and other weather systems. By comparison, the stratosphere is relatively still, dry and virtually cloudless. Supersonic aircraft like Concorde flew in the lower stratosphere, while aircraft like the Boeing 747 fly in the upper troposphere (see Figure 2.2).

Ozone in the atmosphere

Unlike other gases, which are concentrated in the troposphere, about 90 per cent of the ozone occurs in the stratosphere, mainly between about 15 and 25 km in height. Even at its highest concentration it does not exceed 10 ppmv, i.e. only one ozone molecule in every 100 000 molecules of air. If all the ozone in the atmosphere was concentrated at sea level it would form a layer only 3 mm thick. There are about 3000 million tonnes of ozone in the atmosphere, equivalent to about 0.8 tonnes per person on earth. Compared

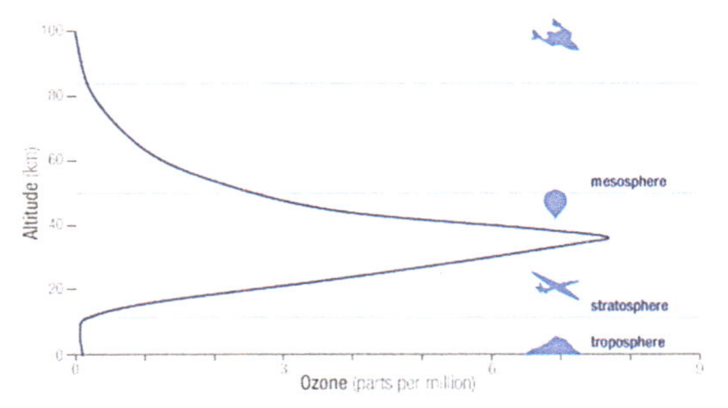

Figure 2.2 Location of the ozone layer
Source: NASA

with the total mass of the atmosphere, the amount of ozone is negligible.

Ozone can be formed by the action of electrical discharges, which is why it is sometimes observed (by odour) near electrical equipment or after a thunderstorm. More frequently, it is formed by the action of ultraviolet radiation on oxygen in the stratosphere. The oxygen atoms in the oxygen molecules split apart, and then recombine with other oxygen molecules to form the tri-atomic ozone (see Figure 2.3).

Because sunlight is essential for the formation of stratospheric ozone, it is mainly formed over the equatorial region, where solar radiation is highest, and from where it is distributed throughout the stratosphere by the slight global wind circulation. Stratospheric ozone levels vary throughout the world, being lowest at the equator and highest towards the poles.

Absorption of ultraviolet radiation

Incoming radiation from the sun is of various wavelengths, ranging from ultra-violet through visible light to infrared.

Ultraviolet (UV) radiation can be very damaging to living organisms, causing sunburn, skin cancer, damage to eyes including cataracts, and premature ageing and wrinkling of the skin. It can also break the food chain by destroying minute organisms such as plankton in the ocean, thereby depriving certain species of their natural food. Plant life and crops can also be devastated by excessive UV radiation.

Fortunately, the damaging forms of UV radiation are absorbed by ozone in the

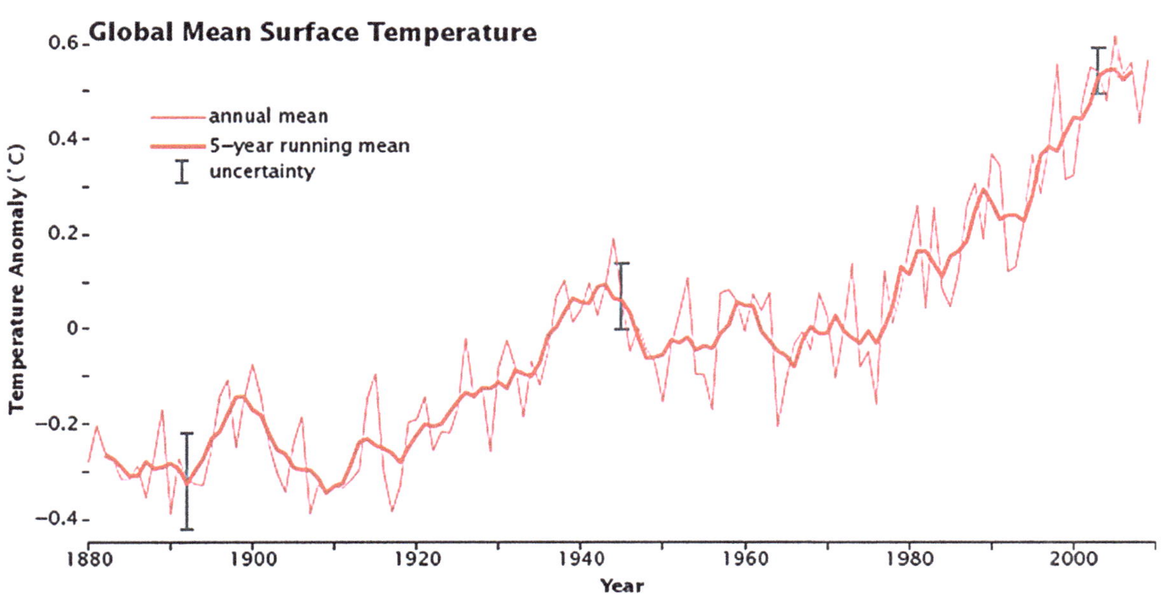

Figure 2.3 Global mean surface temperature
Source: NASA

atmosphere and do not reach the earth. The minute amount of ozone in the atmosphere is sufficient to absorb this radiation. Thus, the ozone layer acts as a giant sunscreen or umbrella enveloping the earth, protecting life from dangerous UV radiation. Depletion of ozone will allow more UV radiation to strike the earth and living organisms.

Another consequence of the absorption of solar energy by ozone is that the upper stratosphere is somewhat warmer than at lower altitudes, which helps to regulate the earth's temperature. Stratospheric ozone absorbs about 3 per cent of incoming solar radiation, thus serving as a heat sink. Loss of ozone decreases the temperature of the stratosphere, which then affects the troposphere and the weather and climate at the earth's surface by increasing the rate of air movement, changing weather patterns and creating violent storms.

Measurement of ozone

Measurement of ozone in the atmosphere began in the 1920s. The main instrument used is the Dobson spectrophotometer, invented by Gordon Dobson, a British meteorologist. Ozone levels are measured in Dobson units (DU). The normal reading for a column of pure ozone at standard pressure and temperature at ground level is 3 mm of ozone, equivalent to 300 DU. Thus, 150 DU means half the normal ozone level.

The ozone hole

In 1956–57, when ozone measurements were started at the British Antarctic base at Halley Bay, levels were found to fall drastically in September and October to 150 DU (i.e. half normal levels). This is also half the levels measured in the northern hemisphere in spring. So surprising were these results that Dobson and the other scientists thought that their instrument was faulty. When levels rose in November to the expected pattern, they saw that the results were correct and that the atmosphere over Antarctica differed from elsewhere.

It has since been established that, during the winter night, the atmosphere over Antarctica is cut off from the rest of the global atmosphere by a wall of westerly winds known as the circumpolar vortex. This vortex surrounds the Antarctic continent like a huge whirlpool and prevents its air from mixing with the rest of the world's atmosphere (see Figure 2.4).

Up to the mid-1970s ozone concentrations over Antarctica within the vortex during winter and early spring remained at about 300 DU and then increased to 400 DU as the vortex broke up in early summer.

From the mid-1970s, the scientists at Halley Bay found the October ozone levels to be declining. Again they disbelieved their figures

Figure 2.4 Antarctic circumpolar vortex

and replaced their instrument, but continued to get the same results. By October 1984 they were certain of their results and published their findings, which showed that spring ozone levels had declined from 300 DU to about 180 DU.

Satellite mapping of ozone levels for the entire Antarctic continent began in the late 1970s. There were consistent results, regardless of how the ozone levels were measured. Ozone levels continued to fall during the Antarctic sunrise, from early September into October.

Throughout the latter part of the 20th century, ozone levels became slowly but progressively lower every spring until, by spring 1987, 50 per cent of the continent had less than 150 DU ozone (50 per cent lower than the normal level). Almost all the ozone in the centre of the depleted area had disappeared (Figure 2.5). Ozone levels at Halley Bay fell by 97 per cent between 15 August and 7 October 1987. The 'hole' occurred at a height of between 17 and 22 km above ground. The ozone loss began earlier than normal and continued until early October – later than normal. (See Figure 2.5.)

From a small beginning, the ozone-depleted area gradually spread during the decade, eventually covering the entire continent.

The term 'ozone hole' refers to the loss of the blocking effect of ozone to UV radiation. The depletion of the ozone barrier has led to the creation of this hole, allowing a much greater amount of UV radiation to reach the earth (see Figure 2.6).

The ozone layer can be likened to an umbrella, which protects the user from rain. An umbrella with holes in it allows the rain through, and the ozone hole is like an umbrella with millions of tiny pinprick holes, each of which allows dangerous UV radiation to come through to the earth.

In spring 1988, the ozone levels had recovered to about 1984 levels. This was not because a solution had been found, but because of a normal two-year cycle in the severity of the hole, with ozone depletion being worse in the odd-numbered years (e.g. 1987) than in the even-numbered years (e.g. 1988). It is believed that this cycle may be due to the reversal of strong high-altitude winds every two years. In the even-numbered years, the

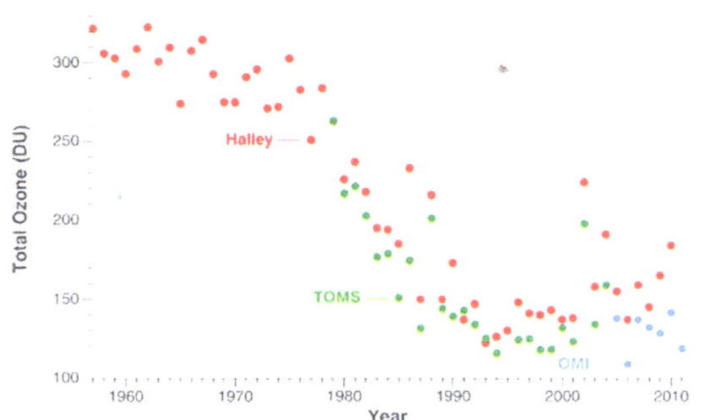

Figure 2.5 Ozone levels at Antarctica
Source: NASA

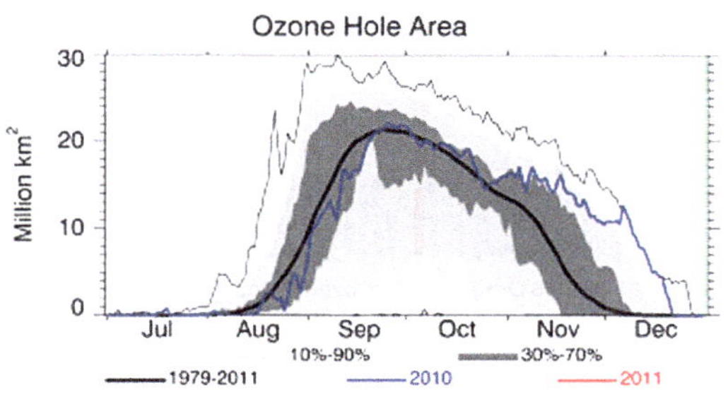

Figure 2.6 Ozone hole
Source: NASA

atmosphere is more turbulent and therefore not as cold, and so the ozone-depleting reactions are less effective. Some scientists hypothesise that the opening and closing of the ozone layer is a naturally occurring phenomenon that has been happening for millions of years. While we have only been able to record ozone levels in Antarctica in the latter half of the twentieth century, it would be hard to argue with the correlation of ozone depletion and increasing chlorine levels in the stratosphere. (See Figure 2.7.)

As described earlier, ozone acts as a heat sink in the stratosphere, making it slightly warmer than it would otherwise be. As ozone was depleted in late 1987, temperatures fell by 8°C – a very significant change. This helped to perpetuate the circumpolar vortex until early December.

With the advent of the Antarctic sunrise in mid-October, the air warms and the rate of ozone destruction slows down. The circumpolar vortex gradually breaks up and the ozone-depleted air trapped with it disperses like tentacles northwards from the polar region. The ozone hole has on some occasions spread to cover Tasmania, Perth, and even as far north as Sydney, as well as reaching Otago in New Zealand and parts of South Africa. This cycle has ebbed and flowed over the years and now, due to the banning of Chlorofluorocarbons (CFCs) (and other ozone-depleting substances), which is considered to be the main cause of ozone depletion, by the Montreal Protocol, the ozone layer may not repair itself until 2050.

The reason why the ozone hole had appeared over Antarctica but did not seem to have occurred to the same extent over the Arctic is because the stratospheric air is far more stable in the southern hemisphere. This is due to the absence of the large land masses and mountain ranges that cause greater mixing of the tropospheric and stratospheric air in the northern hemisphere (see Figure 2.8). The

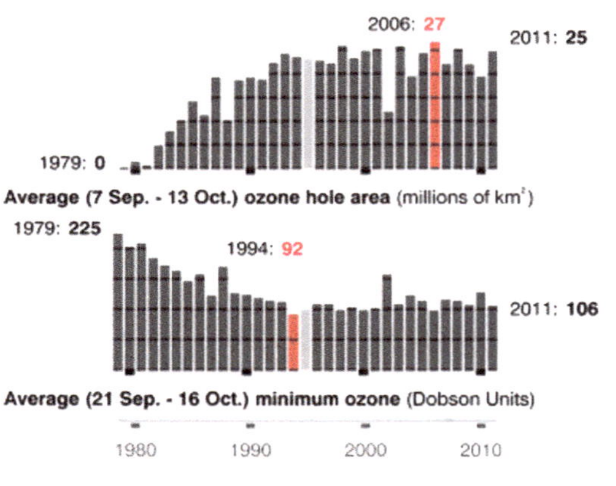

Figure 2.7 Ozone hole change 1980–2010
Source: NASA

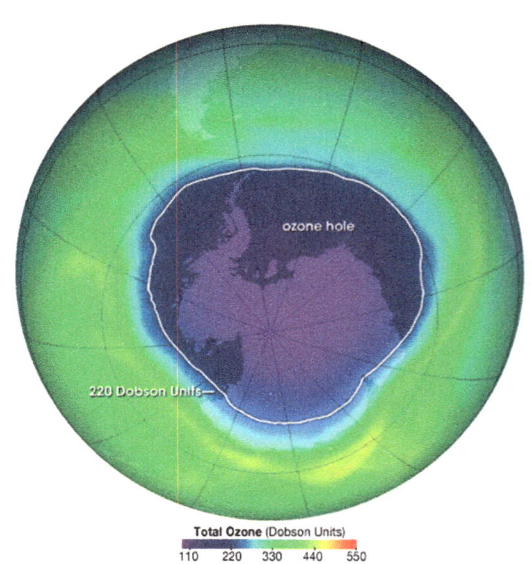

Figure 2.8 Ozone hole and absence of land mass over Antarctica
Source: NASA

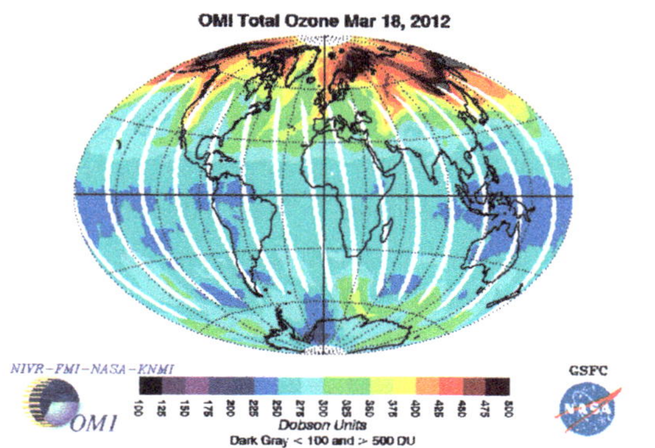

Figure 2.9 World total ozone
Source: NASA

vortex over the Antarctic is far more pronounced than the one over the Arctic and separates the Antarctic air effectively from the rest of the globe's atmosphere during the ultra-cold Antarctic winter. This results in the Antarctic air being some 5–10°C colder than that above the Arctic, and the difference is sufficient to enable chemical reactions to occur, resulting in the loss of ozone. However, the Arctic now also has a significant hole that is fast catching up with the Antarctic ozone hole (see Figure 2.9).

How ozone is being destroyed

Ozone is both created and destroyed by the action of UV radiation on oxygen molecules. Chlorine is the major destroyer of ozone, starting chain reactions in which a single molecule can destroy 100 000 ozone molecules. Such reactions can continue for more than 100 years until the chlorine drifts down into the troposphere or is bound into another compound.

The main sources of chlorine are chlorofluorocarbons (CFCs, also known as freons) and halons. CFCs are artificially made chemicals that were first developed in 1928. They include chlorine, fluorine, carbon and sometimes hydrogen. They are very stable chemicals and are non-flammable, non-irritating, non-explosive, non-corrosive, odourless and relatively low in toxicity. They vaporise at low temperatures, which makes them suitable for use as a coolant in refrigerators and air conditioners, as a solvent for cleaning electronics, in creating bubbles in certain types of foam-blown plastics such as sponge plastic and food packaging, and in cleaning solvents for dry cleaning. Halons are mainly used in fire extinguishers.

Worldwide use of CFCs in 1990 was about 700 00 tonnes. Although aerosol use has declined steadily, non-aerosol use has risen. During the 1960s and 1970s, aerosol use had become widespread because of the stable nature and non-flammability of CFCs. The production of CFCs peaked in 1974 and then declined as their use in aerosols declined. However, with their current increasing use in non-aerosols, the level of production has exceeded earlier levels.

While the industrialised nations are the major consumers of CFCs, developing nations such as China and India, because of their large populations, have enormous potential to use CFCs for refrigeration and other purposes. Australia's consumption on a per capita basis is among the highest in the world, which is a reflection of affluence and the common use of air conditioners.

In 1974, two chemists at the University of California, Mario Molina and Sherwood Rowland, asked the simple question: 'What has happened to the millions of tonnes of CFCs released over the previous four decades?' Their suggestion was the stratosphere. They hypothesised that the chemical stability of CFCs would enable them to reach the stratosphere and be broken apart by the intense UV radiation, thus allowing them to release chlorine by a process known as photolysis. The chlorine would then react with the ozone, causing its depletion.

Thus, it is not the CFCs as such that cause the destruction, but rather the chlorine they release. The research of the British scientists at Halley Bay during the 1980s, together with the international research program in late 1987 in which samples of stratospheric air were obtained by high-altitude flights over Antarctica, proved the link between CFCs and ozone destruction.

A further factor contributing to the loss of ozone is the existence of polar stratospheric clouds that form during the Antarctic winter in the very cold stratospheric air. These are made up of tiny particles of frozen water vapour that condense and form clouds in spring. The clouds act as reservoirs of frozen chlorine during winter that remain until they are thawed in spring, when the chlorine is released and begins once again to react with the ozone over the following five or six weeks, until the vortex breaks up and the stratosphere becomes less stable.

A chlorine atom reacts with an ozone molecule, splitting it apart and attaching itself to one of the oxygen atoms to form chlorine monoxide. A free oxygen atom splits the chlorine monoxide molecule to re-form a molecule of oxygen, and the chlorine atom is then free to attack another ozone molecule (see Figure 2.10).

The CFCs and halons take six to eight years to rise up through the atmosphere. Chlorine as used to disinfect swimming pools is too unstable and breaks down rapidly without rising into the atmosphere. The chlorine needs the CFC molecule to take it up to the stratosphere. The concern is that the current ozone hole and ozone depletion have resulted from CFCs released in the late 1980s; the continuing high levels of CFCs released in the 1980s and the 1990s are only now making their presence felt in the stratosphere.

Effects of loss of ozone on human health

As we have already seen, ozone protects life on earth from damaging UV radiation. It acts as a giant sunscreen, absorbing the UV rays, preventing a certain percentage of them from reaching the earth. Loss of ozone will thus allow more UV radiation to penetrate to the earth, which will adversely affect human health and the environment.

The three areas of our bodies that are adversely affected are the skin, eyes and the immune system.

Exposure of skin to UV radiation can result in sunburn and suntan, and if the exposure continues over a long period, as is the case for outdoor workers, the skin protects itself from UV radiation by gradually thickening and darkening as a pigment called melanin is released into it. Continual exposure of the skin to UV radiation results in its ageing and wrinkling, and increases the risk of skin cancers.

Excessive UV exposure to the eyes will increase the risk of cataracts, which cause cloudiness in the lens of the eye, limiting vision. Other eye problems such as retina damage, tumours on the cornea and 'snow blindness' may also be caused by exposure to increased levels of UV radiation.

The body's immune system protects it from foreign chemicals and infections. If damaged,

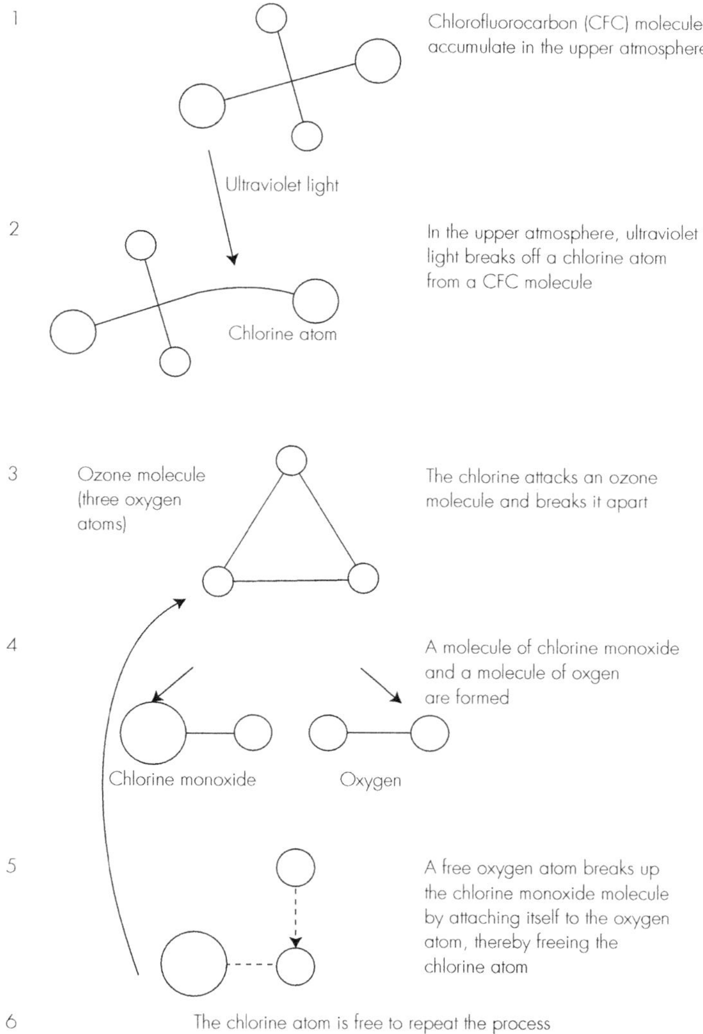

Figure 2.10 How CFCs destroy ozone

the immune system cannot protect the body and infections spread more rapidly. UV radiation reduces the ability of the immune system to reject cancers, although not much is known about why this happens.

Overall, increased UV radiation resulting from ozone depletion has the potential to significantly increase human skin cancers and cataracts, and damage the human immune system. It also adversely affects marine and terrestrial plants and animals. The extent of the damage will depend on the degree by which the earth's ozone layer is depleted. This is estimated at about 2.5 per cent to date. It remains to be seen if we have done enough to control the release of ozone-depleting substances.

Global warming

Global warming is a controversial topic as scientists are still monitoring the temperature changes that are occurring around the world. Is this a natural cycle or are we contributing to the heating of our atmosphere by the greenhouse effect? There appears to be a direct correlation between CO_2 levels and global temperature.

The greenhouse effect

The greenhouse effect is caused by the addition of foreign gases to our atmosphere causing excessive UV radiation to enter the atmosphere and instead of being deflected back into space, the UV rays reflect off the inside of the atmosphere back to the earth's surface. It is a little like getting into your car on a hot day, the heat cannot readily escape. (See Figure 2.11.)

Carbon dioxide (CO_2) and methane are the main greenhouse gases, although there are a number of others including CFCs. By far the major cause of the greenhouse effect is burning fossil fuels, from cars, trucks, power stations and aircraft. They release huge amounts of CO_2 into the atmosphere, more than can be absorbed by the plant life on earth through photosynthesis. Methane is also released into the atmosphere by animals and swamps and wetlands, while there is less methane to contend with it is far more damaging as a greenhouse gas than CO_2.

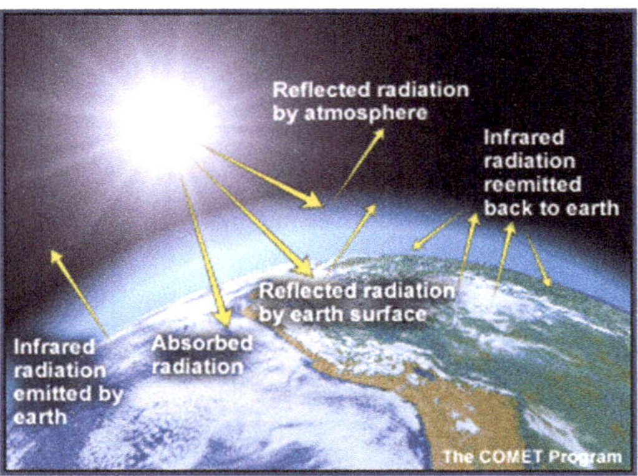

Figure 2.11 The greenhouse effect
Source: ©2000–2001 University Corporation for Atmospheric Research. All Rights Reserved

A very small contribution to global warming may come from the release of air-conditioning refrigerants into the atmosphere through system leaks, service losses and vehicle accidents. All air-conditioning gas in Australia and New Zealand is now required to be recovered and recycled so refrigerant loss has been minimised (see Figure 2.12).

It is important that when new refrigerants are developed their Global Warming Potential (GWP) is kept very low (see Figure 2.13).

New blow to ozone recovery

Since the discovery that emissions of a key hazardous gas have continued to rise instead of levelling off and declining, Australian scientists now expect that the ozone layer will not recover until 2050 (see Table 2.2). Atmospheric tests at Cape Grim in Tasmania have shown that emissions of the fire-retardant gas Halon 1211 are increasing and now stand at about 200 tonnes per year. Concentrations of this gas in the ozone layer have increased by 25 per cent over the past decade, making it responsible for about 20 per cent of the current ozone destruction. Halon is considered the most dangerous of the ozone-depleting substances, now that CFCs have disappeared. It appears that the scientists who created the original models to predict emission rates of ozone-depleting gases had miscalculated the emissions of Halon 1211 from third-world countries.

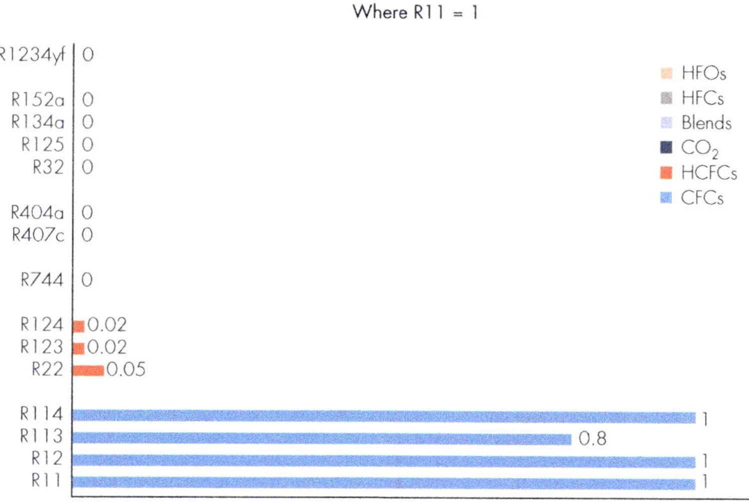

Figure 2.12 The ozone-depleting potential (ODP) of common refrigerants. The main offenders are CFCs and HCFCs, both of which contain chlorine

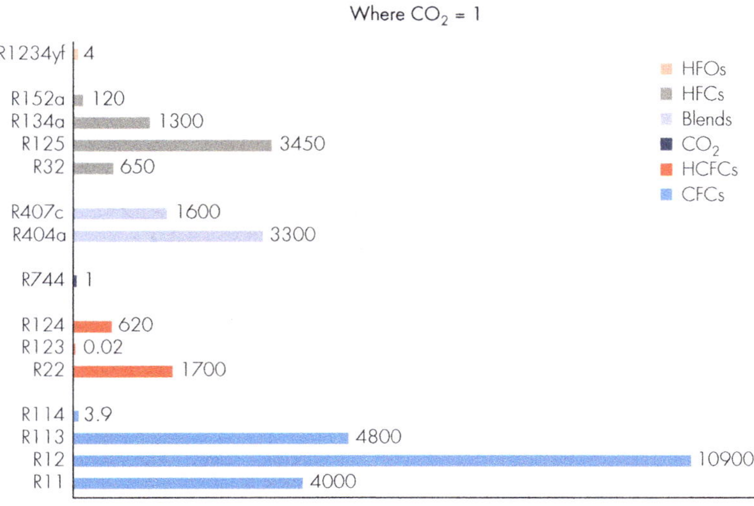

Figure 2.13 Global warming potential (GWP) of common refrigerants. Note the GWP of R12 (no longer used) and the new refrigerants R1234yf and R744

Table 2.2 Atmospheric lifetime of common refrigerants

Refrigerant	Compound	Atmospheric lifetime in years
R1234yf	$C_3H_2F_4$	0.03
R152a	CH_3CHF_2	1.4
R134a	CH_2FCF_3	14
R125	CHF_2CF_3	29
R32	CH_2F_2	4.9
R744	CO_2	>500
R124	$CHClFCF_3$	5.8
R123	$CHCl_2CF_3$	1.3
R22	$CHClF_2$	12
R114	$CClF_2CClF_2$	300
R113	CCl_2FCClF_2	85
R12	CCl_2F_2	100
R11	CCl_3F	45

Sustainability

With nearly 7 billion people on planet Earth and a forecast population of 10 billion by 2050, we need to ensure we can meet the needs of the present generation while protecting the needs of future generations. This is the basis of the concept of sustainability. We need to reduce our consumption of resources by not only recycling materials, but by becoming more efficient in our product design, always keeping the environment in mind.

In Figure 2.14 you can see that social issues and economics are also important when considering sustainability. Interaction between these and environmental concerns is required to achieve an ethical responsibility for caring for our planet.

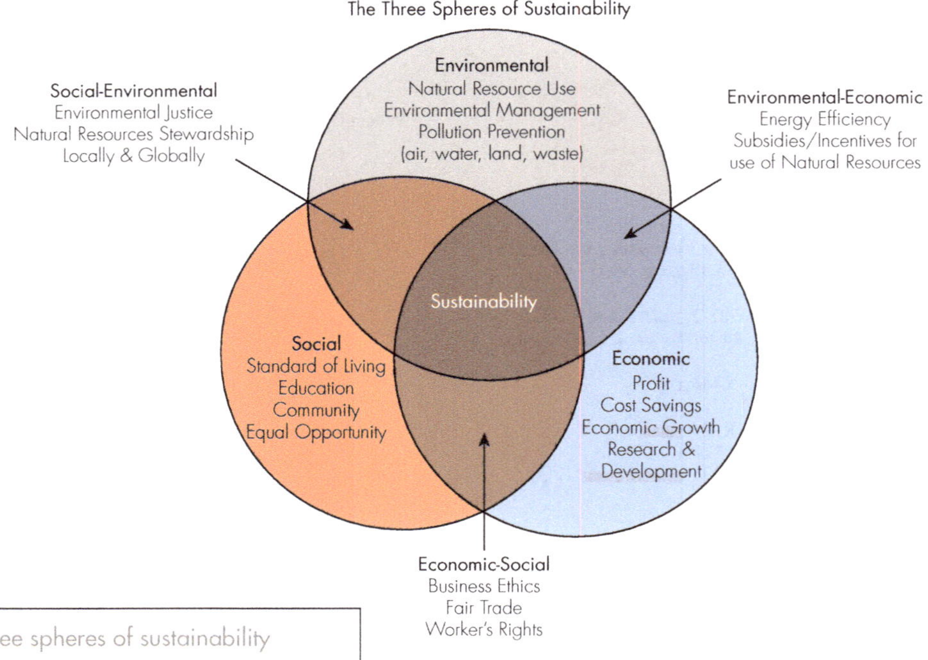

Figure 2.14 The three spheres of sustainability

Chapter-revision questions

EXERCISE

To view up-to-date data and diagrams of the ozone hole go to: http://ozonewatch.gsfc.nasa.gov/. Research the data on this website.

SELF-CHECK QUESTIONS

1 Which gas makes up the largest percentage of earth's atmosphere?
2 Between which two layers of the atmosphere is the ozone layer located?
3 How far, in kilometres, is the ozone layer from earth?
4 What are ozone levels measured in?
5 What is ozone composed of?
6 What is a circumpolar vortex?
7 Where is the ozone hole most prevalent?
8 What is the effect on earth of the depletion of the ozone layer?
9 What chemicals are thought to be responsible for destroying ozone?
10 Which gas is thought to be most responsible for the greenhouse effect?
11 Where is the chemical halon often used?
12 What is the estimated atmospheric lifetime of R134a?
13 Which gas has the lowest atmospheric lifetime?
14 What is the estimated year that the ozone layer will have repaired itself?
15 What are the three main areas to consider when discussing sustainability?

CHAPTER 3

The human body and heat

Objectives

On completion of this chapter the student will be able to:

› state the three methods of heat transfer
› determine the body comfort temperature zone
› understand the term 'humidity'
› state the effects of heat on matter
› define the terms 'sensible' and 'latent' heat
› name the sources of heat load within a motor vehicle
› state the different temperature scales and thermometers used
› define the term 'pressure'
› know how temperature and altitude affect pressure.

This chapter covers some of the essential knowledge and associated skills involved with:

› AURT222670A Service air-conditioning systems – basic heat theory
› AURT202170A Inspect and service cooling systems – pressure and boiling.

The human body and heat transfer

The human body has a normal temperature of 37°C. Although this body temperature may be felt on the skin, a more accurate deep tissue measurement is usually taken under the tongue or under the arm. A good understanding of how the body deals with excess, or insufficient, heat is essential for an air-conditioning student.

How the human body produces heat

The human body converts food into heat. Food is therefore potential heat energy.

Heat energy is measured in joules. You may have noticed the term 'low-joule' on some dietary products. This refers to the energy content.

Food is converted by the body into other compounds, some of which are stored as fats for future use.

Any movement by the body will use up the stored energy and create heat. The more a body exercises, the hotter the body will become, and it must get rid of this excess heat.

How the human body releases heat

There are three ways by which the human body releases heat to the surrounding air:

- **convection**
- **evaporation**
- radiation.

Convection

The human body's ability to lose heat by convection is based on these phenomena:

- Heat always moves from a hot surface to a surface that contains less heat. So, for example, heat will move from a human body to the surrounding air if the temperature of the body's skin is greater than that of the air.
- Heat always rises. An example of this can be seen in the way steam from a boiling kettle always rises.

If we apply this to the way a human body loses heat by convection, then:

- the body will give off heat to the cooler ambient air
- the ambient air temperature will rise as a result
- cooler air will replace the ambient air as it rises, and this air will become warmer. The cycle then continues, creating air movement or a convection current.

Radiation

Radiation occurs when heat is transferred by means of heat rays. Remember that heat will always transfer from a hotter object to a colder one. When you sit in front of an electric heater because you are cold, the heat rays travel through the air and make you feel warmer, yet the air between you and the heater remains constant. These types of heater are often referred to as radiators.

Radiation does not rely on the movement of air as convection does, and the temperature of the ambient air does not affect the transfer of heat.

Evaporation

Evaporation is the process whereby a liquid becomes a vapour when heat is added, losing some of its heat as it does so. When the human body feels hot it tends to perspire through the pores of the skin. As the sweat evaporates, it removes some of the heat. Cool dry air, such as that which comes from an air-conditioning system, increases the evaporation process.

Perspiration indicates that the body is producing too much heat, which can be removed by convection, radiation or normal evaporation.

Factors that may affect human body comfort

There are three conditions that determine whether or not we feel comfortable – temperature, **relative humidity** and air movement.

Temperature

Colder air will increase the rate of convection as it lowers the temperature of surfaces, and so increases air movement. The radiation of any heat will now increase because of the difference in temperature between surrounding surfaces and the body. If the air temperature is warmer, convection air movement will be slower and the radiation rate will decrease. Colder air on the human body will increase the rate of evaporation, while warmer air tends to decrease it. The rate of change depends on how much moisture is in the air, and how much air movement is involved.

Humidity

Humidity is the amount of water vapour in the air. It is usually expressed as a percentage (%). If the relative humidity is described as 75%, it means that the air holds three-quarters of the amount of moisture that it is capable of holding at that particular temperature.

When the relative humidity is high, the air is damp, and evaporation will be slowed down because the air already contains a large amount of moisture. If, however, the air is dry (i.e. relative humidity is low), the evaporation process will readily give up moisture to the surrounding air. An air-conditioning system removes moisture from the interior of a vehicle and lowers the humidity to approximately 48 per cent.

The human body feels most comfortable at somewhere between 22°C and 27°C, and at between 45 per cent and 50 per cent relative humidity. This is called the **human comfort zone**.

Movement of air

The movement of air over the body is important for body comfort.

As air movement increases, body heat is removed faster by evaporation, convection and radiation. An air-conditioning system circulates cooled air around the interior of the vehicle, and this movement in turn helps to remove heat.

Heat and matter

It is important that an air-conditioning technician understands how heat energy affects the various components of an air-conditioning system. The structure of matter, heat and pressure are the main laws of refrigeration covered in this section.

Matter is found in three forms – solid, liquid and gas – and matter is anything that occupies a space and has a mass.

The most common substance on earth is water, which covers 75% of the planet and is a liquid in its natural form (see Figure 3.1). If enough heat were to be added to an amount of water in its liquid state, it would boil and change its state to a gas (vaporise). If enough heat were to be removed from the same amount of water, the water would

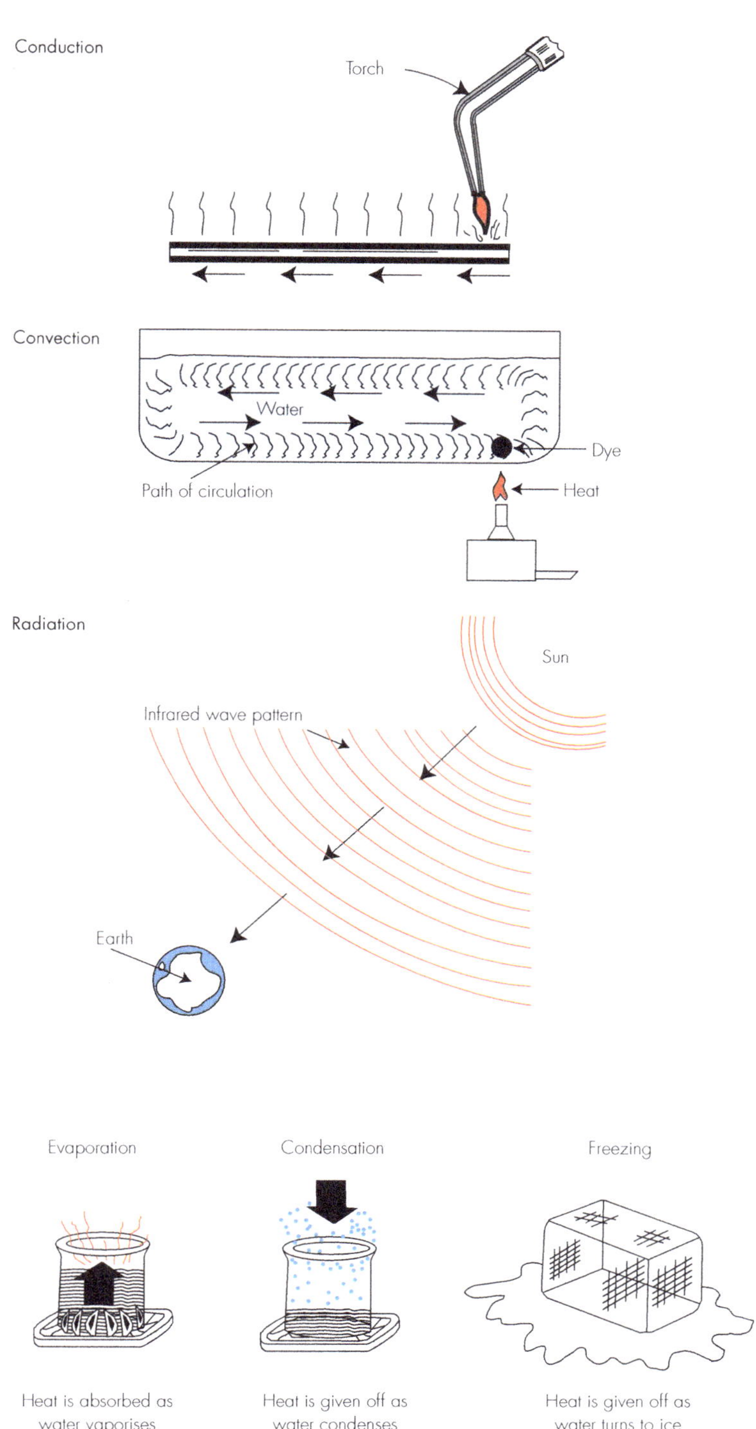

Figure 3.1 Methods whereby heat is gained and lost

freeze and change its state again, but this time it would become a solid (ice). So you can see that we can change the form of water by either removing or adding heat.

Most forms of matter have one natural state, i.e. solid, liquid or gas. However, by adding or removing enough heat, we can transform them from that natural state. Different types of refrigerants react differently when heat is added to an air-conditioning system, and therefore cooling efficiency and pressures can be altered by changing refrigerants.

The structure of matter

If we were to take a grain of sugar and break it down to its smallest particle, we would have a molecule of sugar. We would need an electron microscope to be able to see them, but the resulting sugar molecule is made up of even smaller particles of matter known as atoms – in this case atoms of hydrogen, carbon and oxygen. These atoms stay together to form a molecule of sugar until the temperature of the molecule changes enough to cause the atoms to leave it to form another substance. For example, if we heat some sugar in a saucepan until it melts, and then continue to stir it, we will eventually have caramel. The atoms will have moved to form new molecules because we changed the state of the sugar by applying heat to it.

The movement of molecules

All substances consist of millions of molecules that are the same. What makes one substance different from another is the makeup of its molecule, i.e. the types of atom contained within the molecule.

Whether a material is a solid, liquid or gas, the molecules in it are always moving. This is known as *kinetic energy* ('kinetic' means moving). The molecules in solids only move enough to cause a slight vibration, but if heat is added, a change of state will occur (solid to liquid) and the molecules will move more freely. If we continue to add heat, another change of state occurs and the liquid becomes a gas, with the molecules moving about rapidly (liquid to vapour). This is illustrated in Figure 3.1.

We can, therefore, say that whenever enough heat energy is added or removed, a substance will change its state and there will also be a change in the motion of the molecules. Heat is the key to controlling molecule movement in a substance.

Heat is a form of energy that can be transferred from one place or object to another. However, this transfer cannot take place unless there is a difference in the temperature of the two objects at the time when a heat transfer is about to take place; for example, the difference in temperature in the **condenser** (80°C) of an air-conditioning system from the ambient air passing over the condenser (35°C).

Every molecule contains some heat. As heat is a form of energy, it cannot be created or destroyed, but it can be moved from one place to another, or from one form of energy to another. As already mentioned, heat can travel only from hot to cold, and only by conduction, convection or radiation (see Figure 3.1).

- *Conduction* is the movement of heat through solids.
- *Convection* is the transfer of heat through the movement of a liquid or gas.
- *Radiation* is the transfer of heat through a medium such as air, although the medium does not actually become hot.

Sensible heat

The term 'sensible heat' is used to describe any heat that can be felt or measured. A good example is the heat in the air that surrounds us or an object, known as the *ambient air temperature*. Sensible heat can be measured with a thermometer.

When water exists as a solid, its temperature is below 0°C. As a liquid, its temperature is between 0°C and 100°C, while as a vapour it is above 100°C (see Figure 3.2).

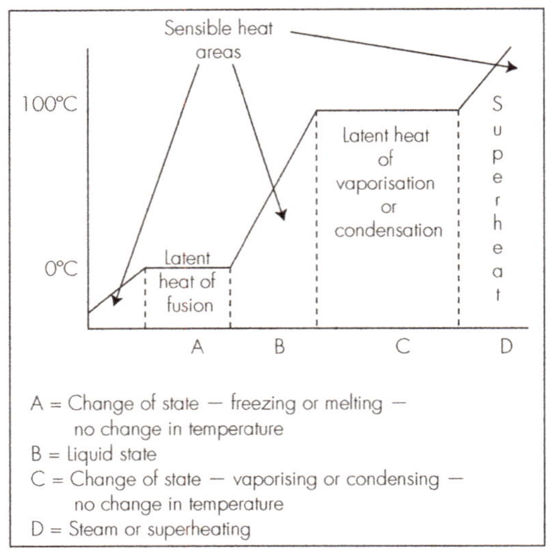

Figure 3.2 Latent heat and sensible heat areas

Measuring heat quantity

The quantity of heat that enters any material is measured in calories. The *gram calorie* is known as the 'small calorie', and is normally used when measuring minute quantities of heat. However, the more common unit that we will discuss is the *kilogram calorie* or 'large calorie'. One kilogram calorie (kg-cal) is the amount of heat needed to raise the temperature of one gram of water 1°C.

Latent heat

The term 'latent heat' is used to describe the quantity of heat required to cause a change of state. This is not sensible heat, as it cannot be felt or recorded on a thermometer. Water at sea level between 0°C and 100°C is called **sub-cooled liquid**, while water at 100°C is known as **saturated liquid**. This means that water at 100°C contains all the heat it can possibly hold, and any more heat would cause a change of state (to a gas).

To cause a change of 1 kg of water at 100°C to 1 kg of steam requires 540 kg-cal. This is called the *latent heat of vaporisation*. If we were to remove 244 kg-cal of heat from steam at 100°C, it would condense back into a liquid. The heat given up in this process is known as the *latent heat of* **condensation**.

If the water is cooled to 0°C and we continue to remove heat from the water, it will change its state to a solid: ice. The temperature will not drop any further until 36 kg-cal of heat is removed, and the liquid has changed completely to ice. This is called the *latent heat of fusion*.

Latent heat of vaporisation and condensation is the basis for all air-conditioning refrigerants and explains how they absorb and give up huge amounts of heat quickly without changing their temperature.

Heat **glide** is the time taken to move through the phase of latent heat. Some **blended refrigerants** separate as they pass through the **evaporator**, causing them to move faster through the evaporator than other component parts without absorbing the correct amount of heat and therefore reducing the performance of the system.

Specific heat

A material's ability to absorb and give off heat is known as its **specific heat**. A scale with fresh water as the standard (1) is used to determine specific heat values. Water is one of the best holders of heat. Most substances require less heat per unit of weight to cause the same increase in temperature. However, there are some exceptions to this: ammonia has a specific heat value of 1.100; and hydrogen 3.410.

Refrigerants have the capacity to absorb and give up large amounts of heat quickly, which makes them ideal for air-conditioning systems.

Table 3.1 shows some specific heat values.

Table 3.1 Specific heat values

Substance	Specific heat value
Hydrogen	3.410
Ammonia	1.100
Water, fresh	1.000
Water, salt	0.940
Petrol	0.700
Alcohol	0.600
R134a liquid	1.113
R134a vapour	1.1708
R12 liquid	1.081
R12 vapour	1.544
1234yf liquid	1.106
1234yf vapour	1.601

Heat load

A type of specific heat that concerns an air-conditioning technician is **heat load**. It is essential that the heat load of a vehicle be taken into account when considering how efficiently an air-conditioning system is operating. It is determined by such variables as the vehicle's colour, how many people are in it – even the glassed area of that will allow extra sunlight into the vehicle.

Factors that influence head load are:

- ambient air
- sunlight
- engine heat
- road heat
- number of passengers
- transmission heat
- exhaust system positioning
- glass area
- colour of vehicle
- miscellaneous sources of heat.

The specific heat values of all materials to be used in the vehicle need to be considered, and the heat load ascertained, when determining the air-conditioning requirements for a specific vehicle. These are then used to select an air-conditioning system that will adequately cope with maximum heat loads.

Cold – the absence of heat

We have already stated that all substances contain some heat – even ice contains a considerable amount of heat, although we would consider ice to be cold. So the question is 'How do we define cold?' In its true sense, the word 'cold' should mean the absence of all heat. There is a point where all molecular movement stops, and this is the point at which there is no heat. This point is known as *absolute cold.*

Absolute cold occurs at a temperature of −273.16°C. All values above this temperature contain some heat, as there is some molecular movement.

Thermometers

Many years ago it was recognised that there was a need for a device that would measure the intensity of heat in an object. In 1585 Galileo Galilei invented such a device, now known as a *thermometer*. His invention was a crude water thermometer.

There are four scales currently used to record the intensity of heat (or temperature). They are described here.

1 The Fahrenheit scale

In 1714, a scientist named Gabriel D. Fahrenheit constructed a thermometer that utilised a column of mercury. His thermometer was to become one of the most widely used measuring devices. Fahrenheit's scale was devised by finding the coldest day in the coldest country on earth. After talking to sailors, Fahrenheit set sail for Iceland. He called the lowest reading that he recorded on his thermometer 'zero', then waited for the ice to melt. He found that his scale now read -^r of its original volume. He also found that the **boiling point** of water was 10^- of the original volume of his mercury column. So Fahrenheit designated 32° as the **freezing point** and 212° as the boiling point of water. On this scale the normal body temperature was 98.6°. The Fahrenheit scale is still used in some countries but is being replaced by the Celsius scale.

°F	°K	°C	°R	
−460	0	−273	0	Absolute cold
32	273	0	492	Water freezes
77	298	25	537	
122	323	50	582	
167	348	75	627	
212	373	100	672	Water boils

Figure 3.3 A comparison of key points on the various temperature scales

2 The Celsius scale

Anders Celsius, a Swedish astronomer, designed the centigrade thermometer. Again, this was based on the freezing and boiling points of water, with 0° being freezing and 100° the boiling point. Celsius is used in the metric system and is by far the most common temperature-measuring scale used in the world today.

3 The Kelvin scale

W. T. Kelvin (Lord Kelvin) designed a scale that is now commonly used in scientific work. He recognised the point at which there is no heat (absolute cold) and started his scale there. Therefore, absolute cold is zero. The scale continues to use the Celsius concept of 100° between the freezing and boiling points of water. Thus, on the Kelvin scale water freezes at 273°K and boils at 373°K (see Figure 3.3).

4 The Rankine scale

W. J. M. Rankine was a Scottish engineer and physicist who designed an absolute temperature scale based on the Fahrenheit scale, where 0° is absolute cold and the freezing point of water is 492°. To convert a temperature on the Rankine scale to Fahrenheit, simply add 460° to the Fahrenheit reading (see Figure 3.3).

Several types of thermometer are available to the service technician, with the digital type being the most accurate and popular (see figures 3.4–6).

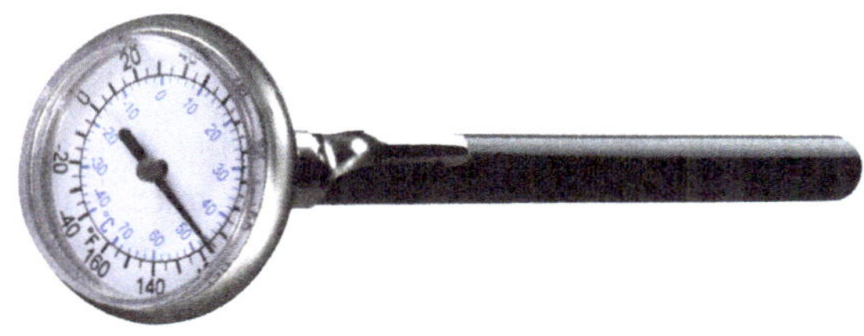

Figure 3.4 Analogue dial thermometer
Source: Mastercool

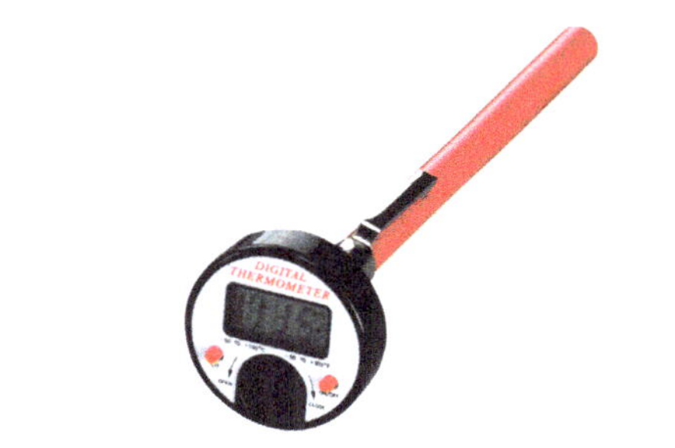

Figure 3.5 Digital pocket thermometer
Source: Mastercool

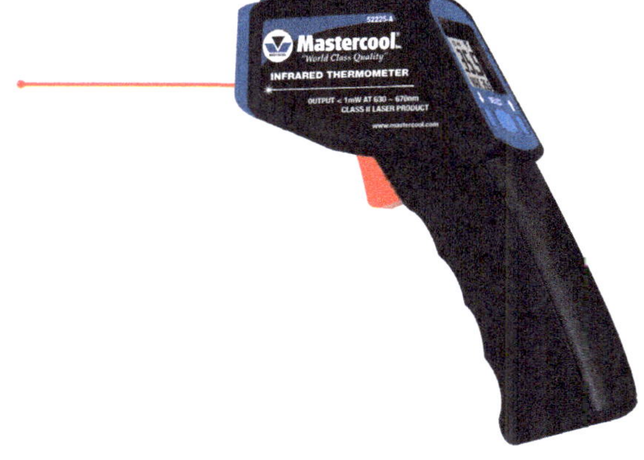

Figure 3.6 Dual temperature digital thermometer, contact and non-contact
Source: Mastercool

The accuracy of these thermometers is usually within 1 per cent throughout the measuring scale. Several ranges are available, covering temperatures from −32°C to 140°C.

Pressure

It is essential to understand the concept of pressure, as well as the gauges and scales that measure it.

A good example of the action of pressure is given by the combination of gases we call the atmosphere. This extends nearly 965 km above the earth and is held in place by its gravitational pull.

Atmospheric pressure

The 965 km belt of gas surrounding the earth exerts a pressure that is measured in kilopascals (kPa), or in pounds per square inch (psi), and is known as *atmospheric pressure* (see Figure 3.7).

Atmospheric pressure when measured at sea level is 101.4 kPa (14.7 psi). The higher one travels in the atmosphere, the lower the pressure becomes.

Pressure measurement

Most manufacturers' service manuals refer to an air-conditioning system's normal pressure in kPa (or psi) as **gauge pressure**.

Gauge pressure *indicates atmospheric pressure as zero* on the dial, but it only reads zero when only atmospheric pressure is acting on it.

However, if a gauge shows 101.4 kPa (14.7 psi) with only atmospheric pressure acting on it, then it is an *absolute pressure gauge*. This type of gauge reads pressure from zero and can measure the amount of atmospheric pressure acting on it. It is a type of barometer used for weather forecasting. (Government legislation may refer to absolute pressures, so it is necessary to understand this pressure.)

If absolute pressure is shown but gauge pressure is required, simply subtract 101.4 kPa (14.7 psi) to obtain the gauge pressure. For example, government legislation may require a minimum pressure of 02 kPa absolute when evacuating an air-conditioning system. The corresponding gauge pressure would be:

02 kPa absolute – 101.4 kPa = –99.4 kPa

Temperature and pressure

Pressures above atmospheric are referred to as positive pressure readings, and all readings below atmospheric pressure are known as **partial vacuum** or negative pressure readings.

On a gauge pressure gauge, pressure above atmospheric is recorded as kPag or **psig** (the g is normally dropped). Pressure below atmospheric is recorded as minus kPa or as inches of mercury (in/Hg).

At sea level the boiling point of water is 100°C. At any higher point the atmospheric pressure is lower, and so the boiling point of water is also lower. The rate at which water boils decreases at the rate of 0.6°C for each 300 metre increase in altitude.

Note that water boils when it contains all the heat that it can for a given condition. Thus, water contains less heat when it boils at a lower temperature, and more when it boils at a higher temperature.

Just as the boiling point of water is affected by a **pressure drop**, a pressure increase will also affect the boiling point of water.

As a result of higher pressure, the vapour **superheats** to a higher temperature.

The cooling system in a car is a good example of an instance where the temperature is increased by increasing the pressure. Some vehicle manufacturers have been able to increase the working pressure of a cooling system to 135 kPa (19 psi), thus increasing the boiling point of the coolant under this pressure. For example, the boiling point raises 2°C for every 10kPa of pressure above atmospheric placed on it (see Figure 3.7).

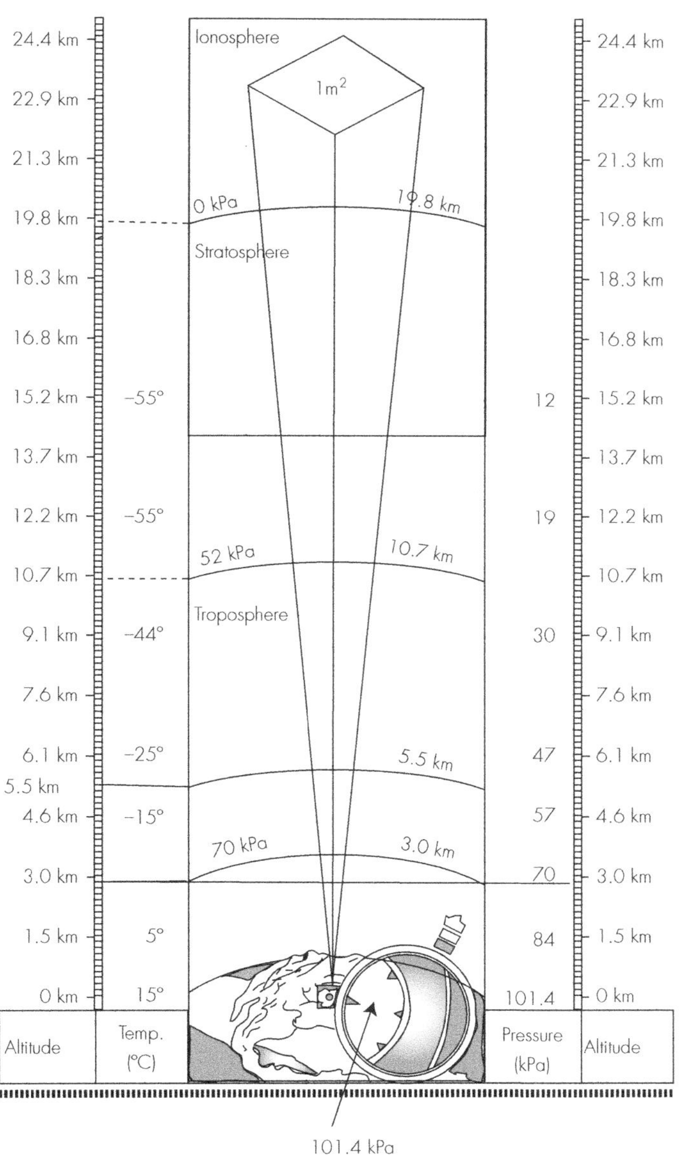

Figure 3.7 Stages of the atmosphere

Chapter-revision questions

PRACTICAL EXERCISE

Turn to Section 2, page 148, where you will find some pressure/temperature problems. By completing these exercises you will develop a greater understanding of the pressure/temperature relationship.

SELF-CHECK QUESTIONS

1 What is the normal temperature of a human body?
2 How is heat energy measured?
3 What are the three ways a human body can release heat?
4 Does heat move from cold to hot or hot to cold?
5 What are three factors that affect human comfort?
6 What is meant by 75 per cent relative humidity?
7 What are the three forms of transferring heat?
8 What is sensible heat?
9 What is latent heat?
10 What are four factors that may affect a vehicle's heat load?
11 Name the four temperature scales.
12 Which temperature scale is used when servicing air conditioners in Australia and New Zealand?
13 What is the atmospheric pressure at sea level?
14 What is the difference between gauge pressure and absolute pressure?
15 At what temperature would a pot of water boil at the summit of Aoraki/Mt Cook (a height of 3754 meters)?

CHAPTER 4

The refrigeration circuit

Objectives

On completion of this chapter the student will be able to:

- state the laws of refrigeration
- describe the composition of refrigerants
- state the boiling point of refrigerants
- use prescribed safety procedures for refrigerants
- use the specific oils for given refrigerant applications
- identify the properties of refrigerant oils.

This chapter covers some of the essential knowledge and associated skills involved with:

- AURT222670A Servicing air conditioning – refrigeration theory

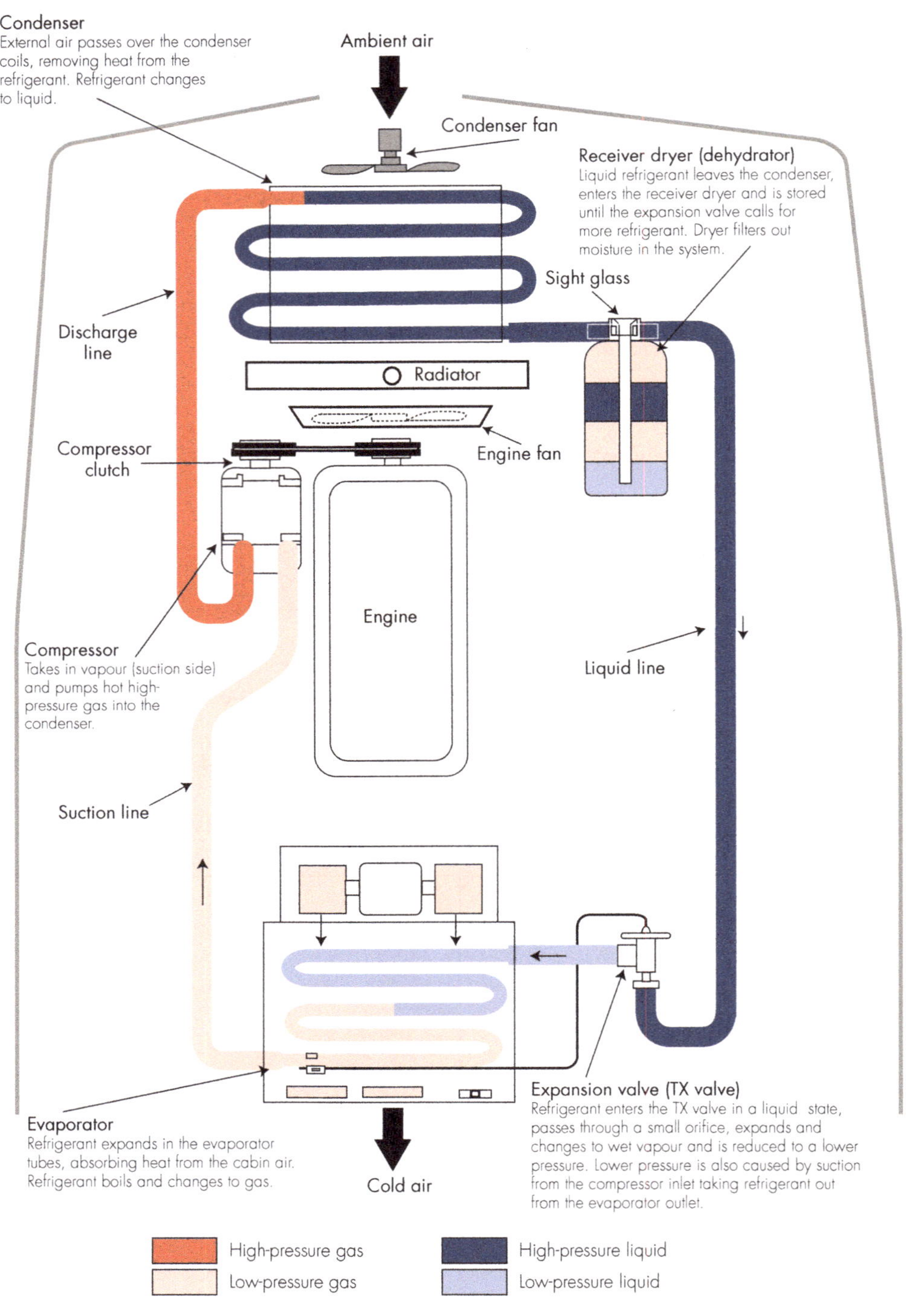

Figure 4.1 (pp. 36–7) Standard TX valve-receiver dryer system
Source: Atkins Carlyle Car Parts

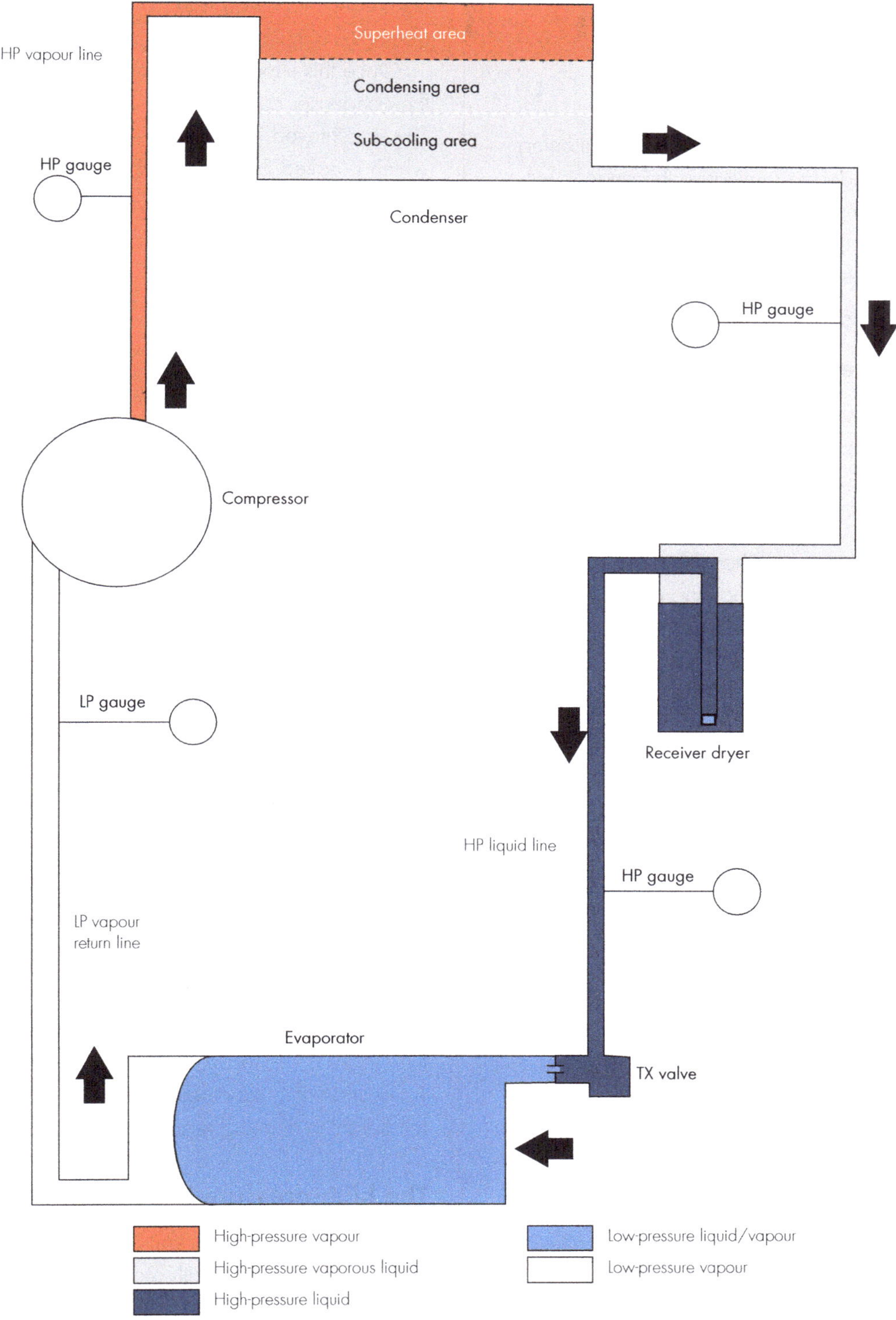
HP vapour line
Superheat area
Condensing area
Sub-cooling area
HP gauge
Condenser
HP gauge
Compressor
LP gauge
Receiver dryer
HP liquid line
HP gauge
LP vapour
return line
Evaporator
TX valve
High-pressure vapour
High-pressure vaporous liquid
High-pressure liquid
Low-pressure liquid/vapour
Low-pressure vapour

System operation

The air-conditioning circuit

The following description of an air-conditioning circuit is intended to familiarise the student with the basic components and their function. The operation of the system does not depend on the type of refrigerant used as the description is the same for all refrigerants. (See Figure 4.1.)

The compressor

The **compressor** receives cold low-pressure refrigerant vapour (approximately 3 to 6°C) from the evaporator. This helps to control the compressor body temperature to a safe level.

- This carries the superheated vapour absorbed from the ambient air passing over the fins of the evaporator.
- This refrigerant is heated further under compression and discharged into the high-pressure vapour line.

These two actions cause the refrigerant to become 'superheated' as it passes on to the condenser. Superheating occurs when the refrigerant is heated to a higher temperature than what is required for it to change state. The **discharge line** transports the superheated vapour refrigerant to the condenser.

Definition superheated – heated to a higher temperature than what is required to change state. Heated further it does not change state but becomes superheated.

The condenser

The refrigerant vapour flows into the condenser, where the superheat must first be removed so that the vapour can be transformed into a liquid.

This is achieved in the top third of the condenser by the ambient air that passes through the condenser, removing the extra heat. This top third part of the condenser is called the 'superheat area'.

Once this superheat has been removed, the condenser can condense the refrigerant. This is achieved in the centre of the condenser, which is called the 'condensing area' (this is the temperature and pressure shown on the high pressure gauge when the system is working). In this area extra heat is removed from the refrigerant allowing it to start to condense (change its state from a gas to a liquid).

Note that in this stage the temperature of the refrigerant is approximately twice the ambient temperature.

The bottom part of the condenser, known as the '**sub-cooling** area', allows the refrigerant to reject extra heat. The temperature of the refrigerant as it leaves the condenser is used to determine the working ability of the condenser. The difference in temperature between the condensing area and the condenser outlet should be at least 8°C.

The refrigerant is now 99 per cent liquid as it passes along the high-pressure line to the **receiver dryer**.

The receiver dryer

The refrigerant flows into the receiver dryer and, if necessary, completes its change of state. It must be 100 per cent liquid before it leaves the receiver dryer. The refrigerant is stored and filtered in the receiver dryer until the TX valve, fixed orifice tube (FOT) and evaporator need it. Both the TX valve and orifice tube need 100 per cent liquid refrigerant to correctly regulate the refrigerant flow.

The high-pressure liquid line

The high-pressure **liquid line** is the line prior to the control valve usually stated as being between the receiver dryer and TX valve.

This is the only area of the circuit in which the refrigerant is a liquid under all normal operating conditions and heat loads.

The expansion valve

The **expansion valve** is also known as the TX valve or FOT.

As the refrigerant flows into the expansion valve it encounters a restriction, which limits the amount that can pass through the valve. This causes the high-pressure liquid refrigerant to become a low-pressure liquid refrigerant before it passes into the evaporator.

The evaporator

The low-pressure liquid refrigerant absorbs the heat as it changes state from the ambient air passing around it and as it moves through the evaporator. As the refrigerant absorbs the heat it begins to change back into a vapour.

By the time the refrigerant reaches the evaporator outlet it should have completely changed into a heat-laden cold vapour. The expansion valve controls the flow of refrigerant into the evaporator, there must be sufficient refrigerant to keep the temperature at the outlet of the evaporator to approximately 0 to 2°C. The refrigerant now travels to the compressor and again begins its cycle again.

Circuit components

The *compressor* is a vapour pump. It is not capable of pumping a liquid, as liquids do not compress. It is designed to draw in the cold low-pressure refrigerant vapour and compress it, raising its pressure and temperature so that it can be condensed by losing heat and changing into a liquid.

The cold refrigerant that it receives keeps the compressor from overheating. This cold refrigerant absorbs some of the heat generated during the compressing process in much the same way as the coolant circulating in the engine cylinder head and block keep the engine at a normal working temperature.

There are three types of compressors; all are discussed in detail in chapter 10:

- reciprocating piston
- scroll
- vane rotary.

The purpose of the condenser is to remove any superheat from the refrigerant vapour and allow the refrigerant to change to a liquid, then pass through to the expansion valve and into the evaporator. This heat is absorbed in the evaporator and the return line by the:

- ambient air passing over the evaporator core fins
- engine bay component heat as it passes through the engine bay on its way to the compressor
- compressor's compressing action.

These three are the sources of the refrigerant's superheat as it enters the condenser. This heat must be removed before condensing can take place. The amount of heat absorbed during the refrigerant largely determines the condenser sizing. This is done in conjunction with the amount of refrigerant within the system and the flow rate through the condenser.

Types of condenser

Tube and fin

This type, shown in Figure 4.2, is made from a single length of round tube with fins attached in order to enable the dissipation of heat. This type of condenser is now obsolete.

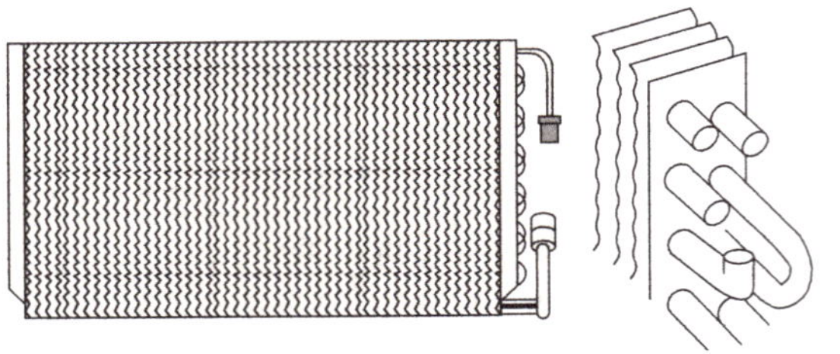

Figure 4.2 Tube and fin condenser
Source: Atkins Carlyle Car Parts

Serpentine or modine

This type, shown in Figure 4.3, uses flat tubing that criss-crosses the condenser, and usually only allows the refrigerant to make a single pass through the condenser.

The modine (high performance) condenser is similar to the serpentine, but has a greater number of passes per centimetre.

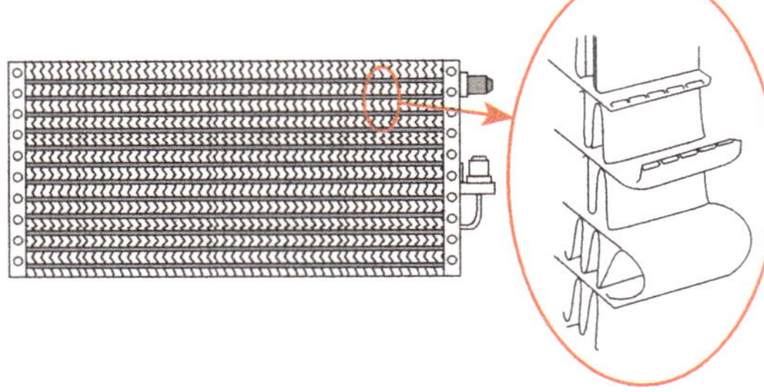

Figure 4.3 Serpentine or modine condenser
Source: Atkins Carlyle Car Parts

Multiflow

This type is also similar to the serpentine, but it allows the refrigerant to flow into multiple sections of the condenser, allowing the condenser to receive it more effectively. This improves the rate at which heat is dispersed.

Parallel flow (high performance)

This type of condenser, shown in Figure 4.4, allows the refrigerant to flow into approximately one-third of the condenser at a time, and then pass across the condenser three or four times before leaving it. Typically:

- 6 fins flow through in the first pass
- 4 fins flow through in the next pass
- 3 fins flow through in the next pass

and 2 flow through in the last pass.

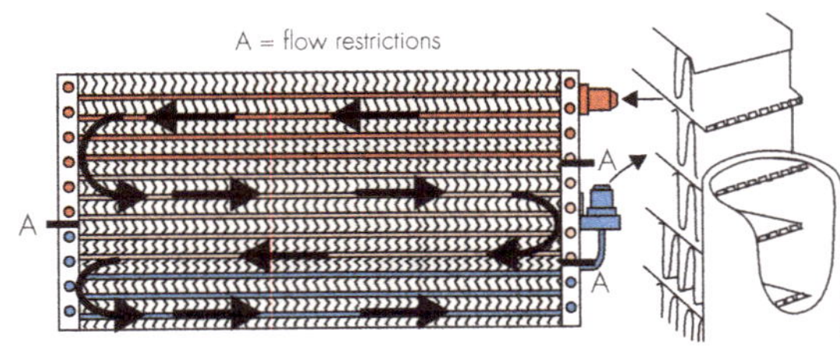

Figure 4.4 Parallel flow condenser
Source: Atkins Carlyle Car Parts

Sub-cooled parallel flow

This is similar to the above condenser, but with one difference: the receiver dryer is attached. This allows extra surface area and therefore better sub-cooling ability.

The last three types of condenser – the multiflow, parallel flow and sub-cooled parallel flow – are the best for **R134a** conversions, as they offer the best heat transfer available. (See Figure 4.5.)

The receiver dryer

This information about the receiver dryer also refers to the **accumulator** that is found on a FOT system. The receiver dryer stores the refrigerant until it is required by the TX valve. It is located between the condenser and the TX valve on the high-pressure side of the system.

The *accumulator*, on the other hand, is fitted between the evaporator and the compressor on the low-pressure return line, and its purpose is twofold, first to make sure all the refrigerant is

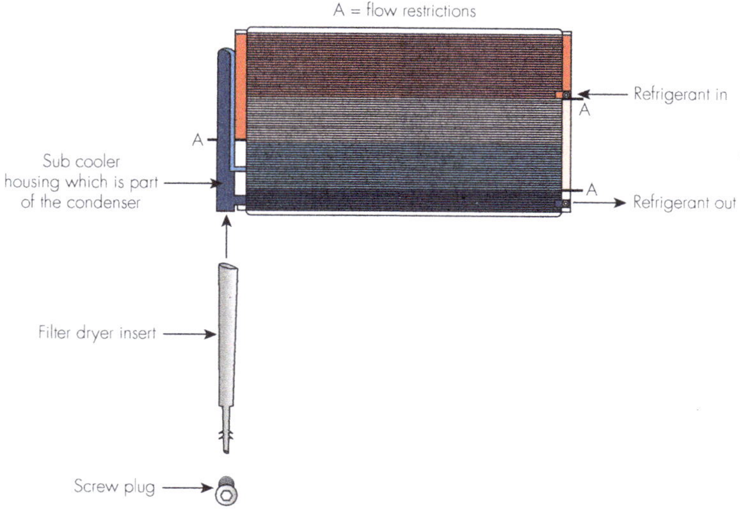

Figure 4.5 Sub-cooled parallel flow condenser

vaporised, and second to store the refrigerant until it is required by the compressor.

Both units store and **filter** particles and moisture out of the system, but that is where the similarity ends. The receiver dryer receives a liquid/vapour high-pressure mix and acts to ensure that it has become 100 per cent liquid before transferring it to the TX valve. The accumulator receives low-pressure vapour/liquid and acts to ensure that it has become 100 per cent vapour when returned to the compressor.

Both units contain a drying agent in a **desiccant** bag, which is designed to remove moisture from the system. Under the Australian Refrigeration Council (ARC) rules, the receiver dryer and accumulator must be replaced at manufacturer-specified times.

The thermostatic expansion valve

The **thermostatic expansion valve** (TX valve) is a pressure-controlling component (**suction pressure**) that is located at, or near, the evaporator inlet. It separates the high- and low-pressure sides of the system, and includes a restriction that limits the amount of refrigerant that can pass through it. The TX valve is able to open and close the size of this restriction in order to allow the amount of refrigerant flow according to vehicle heat load conditions. This is called **modulation**.

As the refrigerant pressure drops, its temperature also drops, thus allowing the refrigerant to vaporise. Depending on their type, refrigerants vaporise at between −26 and −30°C. As the refrigerant passes through the expansion valve and enters the evaporator as a low-pressure liquid, it absorbs heat from the air passing over the fins and vaporises. As a result, air entering the cabin area is cooled.

The fixed orifice tube

The fixed orifice tube (FOT) works in the same way as a TX valve, but cannot modulate, as it is a fixed-size hole set for the average heat load of the vehicle in which it is installed. An after-market FOT has been developed that allows the valve to have two flow rates: one for idle and low speed and the other for higher speed operation, thus altering the suction pressure for better cooling at lower speeds.

The evaporator

The low-pressure liquid refrigerant enters the evaporator and starts absorbing heat from the ambient air around the fins. This now heat-laden liquid starts to boil and vaporise as it passes through the evaporator, until it has completely become a vapour, which is returned to the compressor. Evaporators are usually a single pass unit (see Figure 4.6).

Two of the factors that should be taken into consideration when designing and sizing the evaporator are:

- the amount of air passing over the fins
- the heat load of the vehicle.

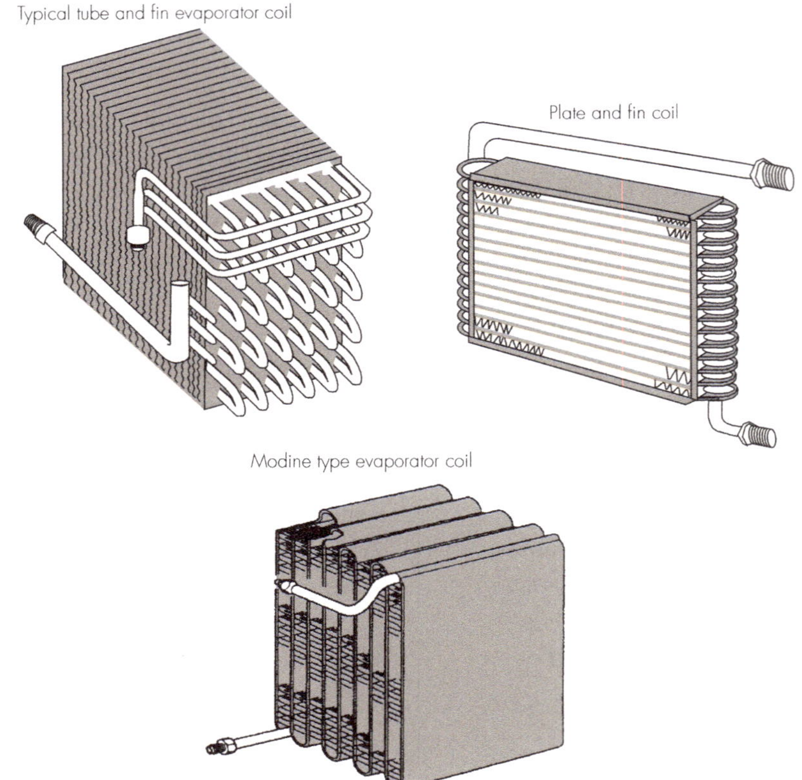

The refrigerant passes through the condenser on several passes. On the second last pass it enters the sub-cooler, has the final pass and leaves the condense

Figure 4.6 Types of evaporator coil
Source: Atkins Carlyle Car Parts

Mufflers

These are usually mounted near the compressor on the low-pressure side of the system. Their purpose is to dampen the noise caused by the flowing refrigerant and the pulses of the compressor.

Electrical devices

The thermostatic switch

The **thermostatic switch** regulates the operation of the compressor. When the temperature of the evaporator drops to a pre-set level, usually between 0 and 4°C, the electrical switch opens to turn the compressor off. This allows the refrigerant temperature to rise, whereupon the switch closes and the compressor starts to pump again.

The high-pressure switch

The **high-pressure switch** is usually fitted between the condenser, and the TX valve on the high-pressure side of the circuit, but can be found on the compressor.

If the pressure rises above a given level, the high-pressure switch disengages the compressor in order to protect the compressor from high-pressure damage. The high-pressure switch is usually set around 3600 kPa.

The low-pressure switch

The **low-pressure switch** is fitted on the low-pressure return line, between the evaporator outlet and the compressor. If the pressure falls below a given point, the switch disconnects the compressor to prevent internal damage due to insufficient refrigerant flow causing a lack of lubrication. The low-pressure switch is usually set at between 30 and 35 kPa.

Combination switches

Some pressure switches combine the above operations and also include a third component of turning on the electrical cooling fans. These types of switches are found on the high-pressure side of the system and operate as follows: high-pressure cut-off above 3600kPa; high-pressure cut-off below 250kPa; fans on above 1400 to 1600kPa. Also, **pressure transducers** (a variable pressure reading device) are now incorporated to do these functions by allowing the computer to operate the system.

The thermal protector switch

The **thermal protector** switch is fitted on the compressor body. If the body of the compressor reaches a given temperature, this switch disconnects the compressor and so prevents internal damage. The temperature level is usually set at or above 115°C.

The surge diode

The **surge diode** is placed in the electrical circuit near the compressor or as part of the relay on an ECU (electronic control units) circuit to protect the vehicle's computer against back **EMF** (electromotive force spike voltage) when the compressor **clutch** disengages. This spike voltage can reach up to 40 to 90 volts and damage the vehicle's computer systems.

Twin evaporator systems

Twin evaporator systems are usually found on vans, people movers and four-wheel drive vehicles (see Figures 4.7 and 4.8).

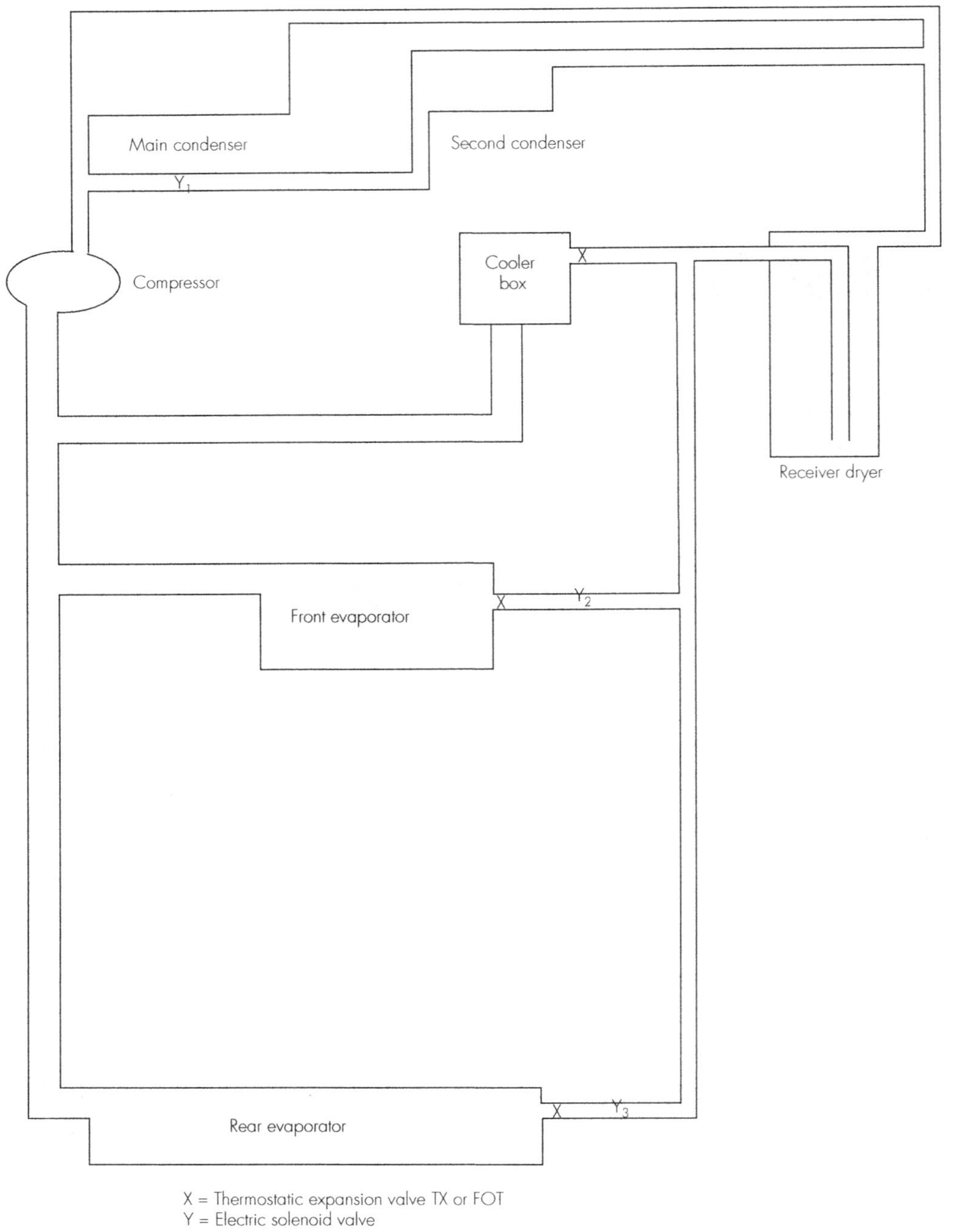

Figure 4.7 Diagrammatic representation of a twin evaporator system

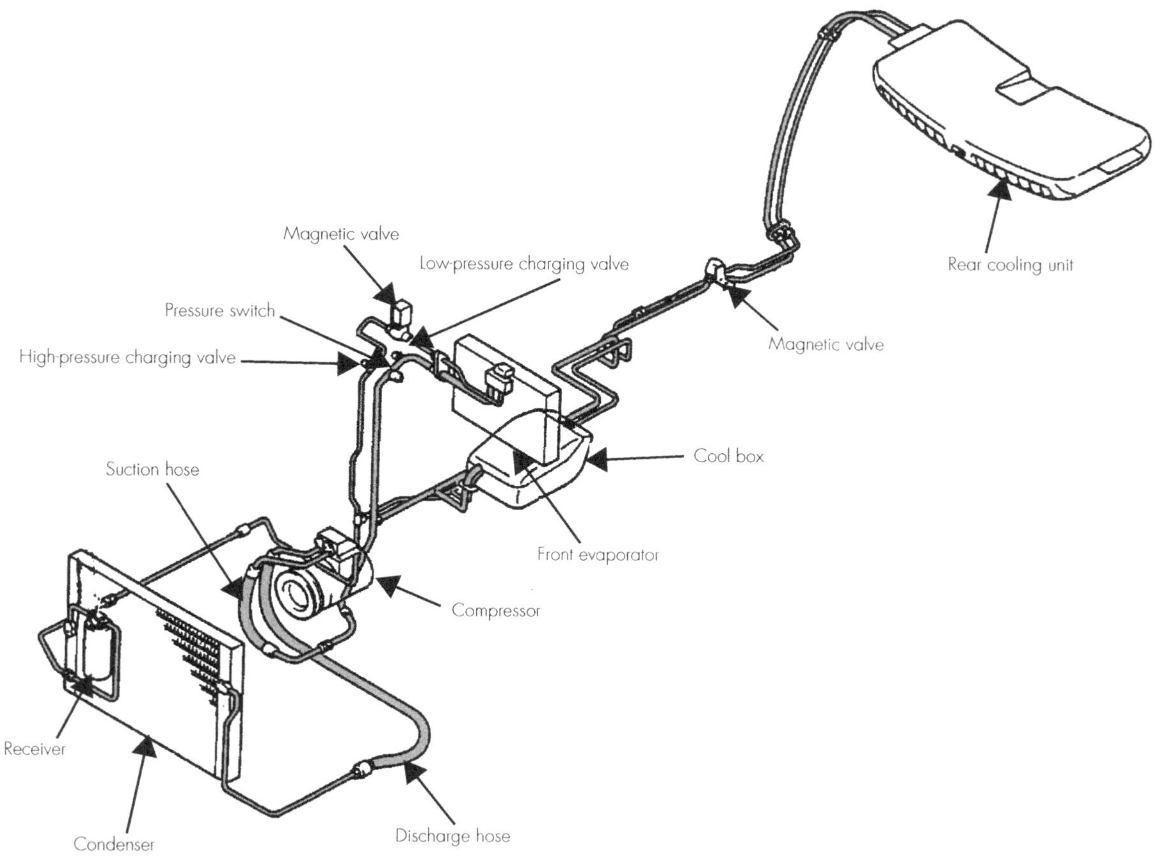

Figure 4.8 A typical twin evaporator system

There can be either one or two condensers in the system. If one large condenser cannot be fitted to the vehicle an auxiliary condenser is fitted in parallel to the first. Two evaporators are also fitted: one for the front of the vehicle and one for the rear.

There can also be a cooler box in the system that will further complicate the refrigerant circuit.

Each evaporator and the cooler box must have a solenoid control valve to isolate that part of the circuit when not in use. The solenoids are identified in Figure 4.7 by the letter Y.

These valves can either be on the high-pressure side of the circuit or in the suction return line. They work in the following manner:

- If only one of the evaporator circuits is operating (either front Y_2 or rear Y_3) then the control valve Y_1 is closed, allowing the main condenser to handle the heat load.
- When both evaporator circuits are operating, then control valve opens, allowing both condensers to handle the heat load.

This type of system must be given close attention when changing from one type of refrigerant to another (**retrofitting**), as the TX flow rate will be too great when all three expansion valves (X on the diagram) are working.

Routine system maintenance

There are several regular checks that need to be performed on a regular basis as part of routine maintenance. Some are noted below.

1 A growing number of new vehicles are fitting a pollen filter into the air intake system. These pollen filters prevent dust, bacteria and other pollen-carrying material from entering the cabin area. They must be inspected regularly and replaced as necessary. If they become dirty they will restrict the air flow into the cabin and lower the capacity of the system to cool and heat. Access is usually from behind the glove box or from the engine area.
2 Air distribution should also be regularly checked to make sure that the airflow is distributed as required. Operate the controls to make sure the air flows to the desired areas.
3 Check that the heater operation for hot air is approximately 50°C and when the heater is turned off, that the air temperature is close to ambient and showing that the heater tap and or flap is closed.
4 Water under the vehicle indicates that the system drainpipe is clear and working. An air-conditioning system removes moisture from the air. If this water remains within the chamber, bacteria may start to grow and a smell will develop.
5 Inspect coolant and air-conditioning hoses for signs of leakage. When air-conditioning hoses and pipes leak refrigerant, oil will also leak out, so oil stains may indicate refrigerant leaks.
6 Operational noises from: (1) fans – the noise is only audible when the fans operate; (2) compressor area – squeals from drive belts may be caused by loose or worn drive belts or worn idler pulley bearings. Worn bearings may also cause noises from the compressor area. If the noise is apparent when the compressor is not operating, the pulley bearings may be worn. Also, if noise is only audible when the compressor is working, the bearings inside the compressor may be faulty.
7 Check and clean the condenser air fins to maintain the correct air flow through the condenser and radiator.

Chapter-revision questions

SELF-CHECK QUESTIONS

1. At what temperature do most refrigerants boil?
2. What is the state of the refrigerant as it enters the condenser?
3. What are the four main types of condenser?
4. What is the function of the receiver dryer?
5. What is the state of the refrigerant as it leaves the evaporator?
6. What is the function of the TX valve?
7. What is the function of the muffler and where is it usually located?
8. Where would you find the high-pressure switch and what is its function?
9. Why is a low-pressure switch used in most systems?
10. At what temperature are thermostatic switches normally set?
11. What is a surge diode?
12. What type of vehicles use a twin evaporator system?
13. Where is the thermal protection switch located?
14. What three functions would a combination switch perform?
15. Where would an accumulator be found?

CHAPTER 5

Refrigerants

Objectives

On completion of this chapter the student will be able to:

› describe and use the more popular refrigerants on the market.

This chapter covers some of the essential knowledge and associated skills involved with:

› AURT222670A Service air-conditioning systems – refrigeration theory

Refrigerant is the term used to describe the compound used in air-conditioning systems to cool the air by removing heat energy from the evaporator to the condenser.

Until 1992 automotive refrigerant was referred to as **R12** as this was the preferred compound. This refrigerant's name is dichlorodifluoro-methane (CCl2F2).

All refrigerants except LPG are man-made. Chemists combine known chemicals in order to produce a refrigerant with the desired results.

R12's instability is the major reason it is not used today.

Refrigerant R134a

NOTE

+ There is no such thing as a 100 per cent drop-in replacement of one refrigerant with another. All refrigerant replacements will require some form of system modification or component replacement, regardless of what any manufacturer of an alternative refrigerant may tell you.

The choice of all original equipment manufacturers (OEMs) for a refrigerant replacement of R12 was Refrigerant R134a (and now **R1234yf** will replace R134a).

The priority was for the refrigerant to have similar characteristics in relation to its pressure, temperature and operating capacity.

The OEMs agreed that R134a (or HFC134a) best met all these requirements, although it was much dearer to produce.

R134a is a pure refrigerant, and will perform to the same cooling levels as R12 when the air-conditioning system is designed or adapted for it. The high-side pressures or **discharge pressures** can be slightly higher than R12 (10 to 15 per cent).

There are alternatives to R134a for older R12 systems but some important questions will need to be addressed:

› Are there any laws regarding the use and installation of the alternative (e.g. the use of HC12 hydrocarbon refrigerant a purified version of LPG)?
› Is additional equipment required? Each refrigerant requires its own equipment to avoid cross-contamination. Two to three per cent cross-contamination will reduce system performance.
› Will the system components be covered by warranties if an alternative refrigerant is used?
› What is the cost of the alternative refrigerant? Some blended refrigerants actually cost more than R134a.
› What is the availability of the alternative refrigerant?
› Does the alternative refrigerant *fractionate* (separate into its various components) as it moves around the system? This fractionating can occur with a number of the blended refrigerants and can result in a loss of system performance. Also, if the refrigerant does fractionate, each single component of the blend tends to leak through the hoses, seals and O-rings at a different rate. For example, R12's molecules are larger than R134a's while R22's are smaller than R134a's so they can leak at a different rate. After a year's use blended refrigerants might not have the same refrigerant constitution as originally intended, and this may affect its performance.
› Does the alternative refrigerant exhibit *glide* characteristics? Glide occurs when the temperature of the refrigerant changes as it changes its state in the evaporator. It causes the evaporator core to warm and lose flow control through the TX valve.

Over the past two decades at least seven refrigerant types have been in common use throughout Australia. These can be found in any state as people travel. Table 5.1 lists these types.

Table 5.1 Common refrigerant types

Refrigerant type	Description
R12	This is a chlorofluorocarbon refrigerant and is ozone-depleting.
R134a	This refrigerant is a hydro-fluorocarbon and is a non-ozone-depleting substance, but it does have a global warming potential. R134a and was the recommended replacement for R12, as it is a pure refrigerant and not a blended compound.

It can be seen in Table 5.2 that two of the blended refrigerants used over the past two decades have a large percentage of the refrigerant is R134a in their compound.

Table 5.2 Blended compounds of two or more base pure refrigerants

Name	Blend components	
FR12	HCFC124	= 39%
	HFC134a	= 59%
	HCR600	= 2%
R401C	HCFC22	= 33%
	HCFC124	= 52%
	HFC152a	= 15%
R413a (Iscecon 49)	R128	= 9%
	HFC134a	= 88%
	HC600a	= 3%
Hydrocarbon (LPG based)	HC600	= 40%
	HC290	= 60%
SP34E (Solpower)	R134a plus Pharmaceutical grade Propane (proportions not available) Ethanol	

Note: if the refrigerant is a 'zoetrope' blend where the number starts with 4 (e.g. R401C) it should be liquid charged, otherwise the refrigerant will fractionate.

The other components of the blend are added into the compound to lower the discharge pressure of the system, approximately to the same level as R12. But, as mentioned above, a blended refrigerant can change the characteristics of the refrigerant under certain conditions, (e.g. high ambient temperatures and/or high humidity conditions leading to **fractionation** and/or glide) and this can affect the performance of the system (i.e. the evaporator temperature does not get as cold).

One blend has a percentage of HCFC22 (R22). This refrigerant has strong anti-plasticiser properties and can affect hoses, O-rings and seals that are not designed for its use, resulting in hardening and crumbling of these parts. R12 systems are rarely designed to come into contact with R22.

All of these blended refrigerants *must* be charged into the system as a liquid so that they will keep the same percentage of each component.

The second last one on the list is an LP gas blend that can become explosive under certain circumstances. It comes under legislation in certain states and is banned in others, but is still legal to use in some states. The last blend manufacturer on the list has not given the percentages of each compound that is used.

LPG hydrocarbon refrigerant FR12

This form of refrigerant is generally used as a replacement for R12 and/or R134a. Its benefits are better evaporator efficiency, and lower pressures and at this stage it is not under any legislation. The amount of this refrigerant placed into the system is usually around 200 g due to its massive expansion rate as it changes from a liquid to a vapour. LPG is not the preferred refrigerant of the majority of OEMs and its long-term effects on the system components are only now starting to appear, e.g. effects on hoses and seals that were not made for that form of refrigerant.

R1234yf (HFO1234yf tetrafluoropylene) CH2 CH CF3

HFO 1234yf is the preferred choice of the European Common Market and a growing number of OEMs around the world. These are some important facts about R1234yf:

- it has an atmospheric life of 11 days (R134a's is 13 years)
- it has a global warming potential of 4 (R134a's GWP is 1430. European laws state this must be less than 150)
- it has a boiling point of −29.8°C (R12 is −29.9°C and R134a is −26.6°C)
- it can be recovered/recycled and recharged similar to R134a
- operating discharge pressures are approximately 100kPa lower than for a similar R134a vehicle under the same ambient temperature and heat load.

Air-conditioning dyes can still be used and with a similar pressure/temperature curve it will mean that only modest changes are required to the system. Therefore **performance testing** and diagnosis testing procedures can be carried over to the new refrigerant. R1234yf has a greater **density** than R134a, requiring a change to the control valves operation (TX valve).

There are no plans to allow R134a systems to be retrofitted to R1234yf due to the flammability factors that make the retrofit impracticable

Leak detectors that meet SAE J2791 standards for identifying a wide range of refrigerants will perform on R1234yf. All safety precautions are the same as all other refrigerants as far as skin contact, eye contact and inhalation. Flame exposure emits a toxic fume and the fire extinguisher to be used is carbon dioxide or dry chemical powder.

New recovery, recycling and recharging equipment will be required to handle this mildly flammable refrigerant. The service hoses will still incorporate the quick fit connectors and they will have different connector profiles to stop their use on R134a systems.

The lubrication oil used in testing has been the **PAG oils** currently used with R134a. Barrier hoses and seals suitable for R134a are used, but the newly designed systems are using seal gaskets instead of O-ring-type seals.

Refrigerant identifiers

With the introduction of so many different refrigerants, cross-contamination is a real possibility. Therefore the use of **refrigerant identifiers** is essential. These units can identify refrigerants types. Identifiers are slower to detect R1234yf and may need to be factory re-set to allow identification. The unit's display shows simultaneously the purity and percentage of air by weight in the system, therefore making them an essential diagnostic tool. These units have an accuracy of 1 per cent over the temperature range of 10 to 49°C.

The refrigerants listed in Table 5.2 are blended compounds of two or more base pure refrigerants and, as such, may exhibit different characteristics from a single pure refrigerant under certain conditions.

Table 5.3 provides the formulation of refrigerants.

Table 5.3 Refrigerant formulation

Type	CFC 12	HFC 134a	HCFC 22	HCFC 124	HCFC 142b	HCFC 152a	HC R600	HC R600a	HC R290	FIC 1311
R12	100%	—	—	—	—	—	—	—	—	—
R134a	—	100%	—	—	—	—	—	—	—	—
R176	15%	—	25%	—	60%	—	—	—	—	—
R401a	—	—	53%	34%	—	13%	—	—	—	—
(MP39)	—	—	53%	34%	—	13%	—	—	—	—
R401b	—	—	61%	28%	—	11%	—	—	—	—
MP66	—	—	61%	28%	—	11%	—	—	—	—
R401c	—	—	33%	52%	—	15%	—	—	—	—
MP52	—	—	33%	52%	—	15%	—	—	—	—
R406a	—	—	55%	—	41%	—	—	4%	—	—
GHG	—	—	55%	—	41%	—	—	4%	—	—
R409a	—	—	60%	25%	15%	—	—	—	—	—
FX-56	—	—	60%	25%	15%	—	—	—	—	—
HC-12a	—	—	—	—	—	—	40%	—	60%	—
OZ-12	—	—	—	—	—	—	40%	—	60%	—
ES112R/ES12R	—	—	—	—	—	40%	—	60%	—	—
Calor gas	—	—	—	—	—	—	40%	—	60%	—
Elgas	—	—	—	—	—	—	40%	—	60%	—
CARE 30	—	—	—	—	—	—	40%	—	60%	—
Freezone (2% oil)	—	79%	—	—	19%					
FRIG C	—	59%	—	39%	—	—	2%	—	—	—
FR-12	—	59%	—	39%	—	—	2%	—	—	—
ADAK-29	—	—	—	—	—	—	—	—	—	—
(ADAK-12)	—	—	—	—	—	—	—	—	—	—
IKON-12	—	—	—	—	—	25%	—	—	—	75%
Hot shot	—	—	55%	24%	18%	—	—	3%	—	—
Kar Kool	—	—	55%	24%	18%	—	—	3%	—	—
ICOR	—	—	55%	24%	18%	—	—	3%	—	—
GHGX4	—	—	51%	28.4%	16.5%	—	—	4%	—	—
AutoFrost	—	—	51%	28.4%	16.5%	—	—	4%	—	—
McCool—Chill-It	—	—	51%	28.4%	16.5%	—	—	4%	—	—
X4	—	—	51%	28.4%	16.5%	—	—	4%	—	—
GHGHP	—	—	—	—	—	—	—	—	—	—

Source: Automotive Electrical and Air Conditioning News

New refrigerant

Refrigerant R134a was the most tested refrigerant as of 1992, but because of its affect on global warming, its lifespan was limited. Over the past two decades advances have been made in system design to lower the leak rate and the amount of refrigerant required in the system, but in January 2011 the European Common Market banned the use of R134a in all new production vehicles. Therefore a new refrigerant had to be found. There were two candidates: CO_2 (R744) and R1234yf. CO_2 is now not the preferred refrigerant type due to its pressure (up to HP 14 000kPa) and its lack of performance at higher ambient temperatures (>35°C).

Refrigerant oils

All moving parts of the compressor including the compressor seal must be lubricated to prevent damage or wear. Refrigerant oil also circulates with the refrigerant. Oil can be found in all the components of the air-conditioning system.

Mineral-based oil

Refrigerant R12 used a mineral-based oil specifically designed for use in that refrigerant's air-conditioning system. This oil should not be used with any other refrigerant as it may not be compatible.

Air-conditioning oil is hygroscopic, i.e. it attracts moisture, which damages the air-conditioning system so oil containers must be kept sealed.

Synthetic refrigerant oils

Polyalkylene (PAG) oils

These oils are preferred by 95 per cent of OEMs for R134a, and different viscosity ratings are used for different compressor types. These are listed in Table 5.4.

Table 5.4 Synthetic refrigerant oils

Synthetic refrigerant oil	Description
PAG 1 or 48	Low viscosity oil
	Diesel Kiki piston type
	Nippon Denso piston type ND 3
	The numbers change with different manufacturers
PAG 11 or 150	High viscosity oil
	Diesel Kiki vane type
	Nippon Denso vane type ND 9
PAG SP 10	Sanden TR & SBD type
PAG SP 20	Sanden SD type

Note: that PAG 1, sp10, PAG 48, and ND8 are all low viscosity oils and PAG 11, sp 20, PAG 150 and ND 9 are higher viscosity oils.

PAG-based oils are not tolerant of the residue mineral oil left in the system from a retrofit. PAG-based oils are also hygroscopic, and are the most costly of the synthetic oils. They can be aggressive when they come into contact with the skin; prolonged contact causing irritation.

Ester-based oils (SW100)

The **ester-based oils** are tolerant of any residual mineral oil that may be left in the system. This oil type is, therefore, usually the choice for a retrofitted system (a system where R12 is replaced by an ozone-friendly refrigerant) or where all the mineral oil has not been removed.

These oils are hygroscopic but cost less than PAG oils. One oil type can be used for all compressors (SW100).

Roc Oil 68 (hydrocarbon-based)

Roc Oil 68 is the newest oil on the market. Preliminary test results of this oil are very favourable, showing:

- improved lubrication
- lower operating noise on certain compressors
- non hygroscopic action.

It is claimed that this oil will suspend any mineral oil left in the system and will therefore not affect the operation or performance of the system.

Whichever oil is selected for the system, manufacturers recommend that a full system charge should be used. If the manufacturer's specifications are not available, the data in Table SP5.2 (pp. 197–8) can be used, but manufacturers' recommendations should always be followed in preference to this guide.

NOTE

+ Compressor oil charge charts are found in Section 3.

Chapter-revision questions

SELF-CHECK QUESTIONS

1. Why is it important to use a refrigerant identifier?
2. What are the two new refrigerants considered suitable for use in new vehicles?
3. What is the main disadvantage in using a blended refrigerant?
4. What is the atmospheric lifetime of R1234yf?
5. What refrigeration lubricant will be used with R1234yf?
6. Is there any difference in operating pressures between R134a and R1234yf?
7. What are some disadvantages of PAG oils?
8. What are the main barriers to retrofitting from R134a to R1234yf?
9. What is the recommended fire extinguisher to be used on R1234yf?
10. Why is oil required to be added to the refrigerant?

CHAPTER 6

Retrofitting – using a different refrigerant

Objectives

On completion of this chapter the student will be able to:

- evaluate condenser cooling
- know if the condenser is actually condensing the refrigerant
- understand the workings of expansion valves
- understand the functions of flexible hoses, O-rings, compressors, receiver dryers and other components
- perform a full retrofit process.

This chapter covers some of the essential knowledge and associated skills involved with:

- AURT322666A Repair/retrofit air-conditioning systems

NOTE

+ The refrigerant used in a particular system is designed to work the best in that system and if that refrigerant is available it should be used in that system.

Most air-conditioning systems have been retrofitted in the past from R12 to R134a, but retrofitting actually involves making a system work satisfactorily using any different refrigerant. For example, it might involve a change from R134a to a blended refrigerant or the reverse, or even R134a to R1234yf.

Evaluation of condensers/ evaporators (heat exchangers)

It may be necessary to conduct an extensive pressure/temperature analysis in order to evaluate the performance of both the condenser and the evaporator on the original refrigerant before it can be considered appropriate to convert the system to another refrigerant. At the least they should be thoroughly **flushed** and cleaned prior to retrofitting to ensure maximum performance.

Usually the components will perform adequately on a different refrigerant, but it is essential that the technician becomes competent in system testing and be able to recognise the various danger signs, and when there are excessive pressures within the system – both low pressure (flow control valve) and high pressure (condenser action).

The system may perform adequately on a cool to mild day, but it may not be able to cope with the increased heat loads placed on it on a hot and humid day. This is the reason why vehicles that have been retrofitted on a cool, mild day can fail to cope on the first hot day, with either the high-pressure switch or the thermal cut-out switch preventing good overall system performance and proper **heat exchanger** (condenser) performance.

The evaporator is also affected by the pressures that develop within the system, and this excess pressure can affect the TX valve's ability to control the flow of refrigerant through the evaporator.

There are two tests to perform on the condenser: pressure and temperature.

Pressure

Is the system working at the correct pressure for the ambient temperature? For example, if the ambient air temperature is 25°C, the air passing over the condenser and the evaporator are approximately 25°C in both cases. With the air conditioning working, the air coming out of the centre vent of the dash will have a temperature of about 3 to 8°C, but inside the evaporator core the air temperature will be about 0°C. This means that the refrigerant has absorbed the heat load of approximately 25°C. Take the ambient temperature and divide it by 2, then multiply it by 100. For example, 25 ÷ 2 = 12.5, now × 100 = 1250 kPa (R134a systems). This should be approximately the high side pressure for 25°C. This pressure can be within 150 kPa due to system design and still be within specification. Note that R1234yf would have a further drop of 100 kPa, making the high pressure 1150 kPa ± 150 kPa.

Another way of calculating this operating pressure for R134a is to double the ambient temperature at the front of the vehicle being tested. This should be the approximate pressure/temperature of the operating system (see Table 6.1).

Table 6.1 System operating pressure/temperature chart

Ambient temp. (°C)	R134a Head pressure	R1234yf Head pressure	Centre vent air temp. (°C)
15	700–850	675–800	0–4
16	700–850	675–825	0–4
17	750–900	725–900	1–4
18	800–1000	775–1000	1–5
19	900–1100	850–1050	2–6
20	1050–1200	950–1100	2–7
21	1050–1225	950–1125	3–8
22	1075–1250	975–1150	3–8
23	1100–1300	1000–1175	4–8
24	1100–1350	1000–1200	4–8
25	1130–1400	1030–1240	4–9
26	1150–1450	1050–1280	5–9
27	1200–1500	1075–1300	5–10
28	1225–1550	1125–1350	5–10
29	1275–1600	1175–1420	6–11
30	1350–1650	1200–1460	6–12
31	1400–1650	1250–1480	6–12
32	1450–1750	1275–1480	7–13
33	1500–1800	1300–1500	7–14
34	1550–1850	1375–1550	8–14
35	1600–1950	1400–1550	9–15
36	1650–2000	1425–1600	10–15
37	1700–2100	1475–1650	10–15
38	1750–2150	1500–1650	11–15
39	1800–2200	1550–1700	11–16
40	1800–2300	1600–1750	12–16
41	1900–2400	1600–1775	12–16
42	2000–2450	1650–1800	12–17
43	2100–2550	1700–1900	13–17
44	2200–2700	1800–2000	13–17
45	2300–2800	1900–2150	13–18

Systems designed for R134a can be on the lower end of the R134a temperature scale, while converted systems will be on the higher side of the scale and can be above the pressures shown for a given ambient temperature. The important thing is to make sure that there is no high-pressure gauge creep, i.e. that the system is handling the heat load of the system.

Note: this chart is to be used as a guide only. The actual pressures may vary according to the particular system design.
All pressures are shown in kPa.

Table 6.2 A pressure/temperature comparison between R134a and R1234yf

R134a kPa	R1234yf kPa	°C	R134a kPa	R1234yf kPa	°C
10	27.4	−24	624.8	604.9	28
20.3	38.1	−22	668.2	643.3	30
31.4	49.5	−20	713.4	683.2	32
43.2	61.6	−18	760.6	724.7	34
55.9	74.5	−16	809.8	767.8	36
69.4	88.1	−14	861.2	812.6	38
83.8	102.5	−12	914.6	859	40
99.1	117.7	−10	970.3	907.1	42
115.4	133.8	−8	1028.3	957	44
132.7	150.7	−6	1088.5	1008.7	46
151.1	168.6	−4	1151.2	4062.9	48
170.6	187.4	−2	1216.3	1117.6	50
191.2	207.1	0	1283.9	1174.9	52
213	227.9	2	1354.1	1234.1	54
236	249.8	4	1226.5	1295.4	56
260.2	272.7	6	1502.4	1358.7	58
285.8	296.7	8	1580.7	1424.1	60
312.8	321.8	10	1661.9	1491.6	62
341.2	348.1	12	1789.1	1597	65
371	375.6	14	2016.6	1783.9	70
402.4	404.4	16	2263.0	1965.5	75
435.2	434.4	18	2532.6	2202.6	80
469.7	465.7	20	2824.3	2435.9	85
505.9	458.4	22	3140.7	2686.4	90
543.8	532.5	24	3484.5	2954.7	95
583.4	567.9	26			

The shaded area indicates ideal low-pressure readings under normal ambient temperatures.

Source: Atkins Carlyle Car Parts

If the ambient temperature is 30°C, then, using Table 6.2, if we double 30°C to get 60°C, the pressure for 60°C is 1424 to 1580 kPa. This method can also be used for diagnosing system operation. R1234yf would be 100 to 150 kPa less than this figure.

If these pressures are maintained, the system and all components are working wel .

Temperature

Is there evidence that the condenser is actually condensing the refrigerant? The discharge or

high-pressure gauge reading shows the actual pressure/temperature within the condensing area of the condenser. If a temperature probe were to be placed on the condenser inlet, it would show a much higher temperature than is indicated by the pressure gauge. This is the superheat load of the refrigerant that is absorbed as it passes through the evaporator and as it passes the hot engine components. Also the pumping action of the compressor increases its temperature yet again. This is what is shown when you place a temperature probe at the condenser inlet. The condenser must get rid of this superheat load before it can start condensing. Thus, a high superheat load can reduce the efficiency of the condenser. On some vehicles this is a problem and is dealt with in the following ways:

- shielding or covering the suction hose or pipe
- shielding the compressor
- re-routing the suction hose or pipe.

The actual condenser condensing efficiency should be known before a retrofit is undertaken. This can be discovered very easily by obtaining the following information from a performance test:

- take a high-pressure gauge reading in °C. (Refer to Figure 6.1 for the equivalent temperature of this pressure.) The temperature reading can be read directly off the high-pressure gauge.
- measure the temperature of the refrigerant at the condenser outlet in °C.

A properly working condenser *must* have a temperature drop between 5 and 10°C.

Note that, if the receiver dryer is attached at the condenser outlet, it can be considered to be part of the condenser (sub-cooled condenser). The temperature should be measured at the receiver dryer outlet and the drop should be between 7 and 12°C.

For example, the high-side pressure might be 1650 kPa, equalling a temperature of 66°C. The temperature at the condenser outlet is 58°C, giving a temperature difference of 66 – 58 = 8°C. This is between 5 and 10°C, so the system is working satisfactorily.

After analysing the above two tests the student should be able to determine the suitability of the condenser to handle the extra loads that a retrofit can place on it.

Evaporators normally prove suitable for a retrofit. No replacement or adjustment to the evaporator is necessary, but the TX valve or expansion valve may be a different matter.

Expansion valves

The TX (thermostatic expansion) valve or FOT (fixed orifice tube) controls the refrigerant flow into the evaporator coils in relation to the heat load placed on the evaporator coil itself. This basic flow is controlled by the superheat **capillary tube** and by pressure **equalisation** passages within the TX valve. Together, these control the opening and closing of the valve in order to maintain the superheat value across the evaporator coil (the amount or flow of refrigerant through the evaporator).

Changing to a new refrigerant will not upset this function. However, the capillary tube of the TX valve has become accustomed to the characteristics of a particular type of refrigerant, and now senses a different refrigerant's characteristics. This means that the TX valve may allow more or less refrigerant to flow for a given heat load, but the valve may also be able to correct this itself as the refrigerant becomes cold, thus compensating for the change of refrigerants.

There is, however, a second important aspect of the TX valve: its sizing (or base flow rate). An R12 valve will have a flow rating increase of about 20 per cent on R134a, because of the lower density/viscosity that is characteristic of the R134a refrigerant, while R1234yf will only be 10 per cent higher.

This may cause two problems:

- On initial start-ups, the TX valve will adopt a nearly full open position, causing a rise in the basic flow rate. This could cause liquid to flood back to the compressor before the TX valve has had a chance to respond to the superheat across the coil and start to close down. This is unlikely to occur in the normal air-conditioned vehicle because of the system design and routing of the **suction line**. But with 'on-roof' mounted units, the risk of liquid **flooding** back to the compressor is greatly increased if the flow of the TX valve is excessive.
- Liquid migration is also a potential problem when the flow rate of the FOT is too high. Liquid migration occurs when the evaporator is flooded (i.e the evaporator fills with liquid) causing it to migrate down the suction line towards the compressor. This problem becomes apparent when there is excessive icing on the return line to the compressor. It is a warning that the flow rate of the FOT may be too high to be used when retrofitting.

In either of the above situations, the TX valve/FOT should be changed for the protection of the system.

If the system has twin evaporators, both TX valves will need to be replaced. The increased flow rate, coupled with two TX valves, will reduce the system's performance.

The system may work normally with either the front or rear evaporator connected to the compressor only, but when the two evaporators are working, the suction pressure will rise and so lower the system's performance.

In the case of the orifice tubes, all such tubes are colour-coded according to size and corresponding flow rates.

Examples of typical sizing and colour are as follows:

TX 2611 orifice tube white	0.072"
TX 2616 orifice tube blue	0.067"
TX 2619 orifice tube red	0.062"

Note that FOT units are sized in imperial units.

In a system that is working correctly and has stabilised, the temperature at the beginning of the evaporator is approximately –8°C. Then, as the refrigerant passes through the first third of the evaporator core, it takes on the heat of the air passing over it without completely changing state. Its temperature remains at approximately –8°C. In the middle of the evaporator (where the thermostat probe is set into or over the evaporator fins) the refrigerant has absorbed as much heat as it can without completely changing state. Now the refrigerant's temperature starts to rise as it completely changes state into a vapour. If the TX valve has sufficient flow, the temperature of the refrigerant as it reaches the evaporator outlet should be between 0 and 2°C and it will be 100 per cent vapour. If the refrigerant flow is low (starved), the temperature of the suction line at the evaporator outlet will be above 3°C, and could be as high as 10°C. If the flow is excessive (flooded), the refrigerant will not completely change its state in the evaporator and the temperature at the evaporator outlet will be below 0°C. A liquid/vapour mix may be flowing down the suction line, still trying to change its state (i.e. vaporise).

Ideally, the refrigerant should completely change its state just before the end of the evaporator.

Flexible hoses and O-rings

R134a has a smaller molecular diameter than R12, which further emphasises the importance of system integrity.

Hoses and seals with low permeability rates (lower leak rates) had to be used on R134a systems.

R134a systems used nylon barrier hoses to lower the leak rate of the refrigerant.

The R12 hose materials are not ideal for R134a systems, but can be used in a retrofit if necessary. This is largely because of the mineral oil and chlorine residue left from the R12 gas, which effectively coats the hoses and seals, helping to seal the pores of the compound and so prevent excessive leak rates of the R134a refrigerant. The decision to change hoses usually depends on:

› the condition and age of the hoses and their joints (loose fittings, clamps or swaged joints)
› the length of the hose run, as long hose runs are generally dryer and therefore prone to higher leakage rates.

Replacing hoses is a cost-effective way of ensuring that the system maintains its integrity. All disturbed O-rings *must* be replaced.

It is recommended that all O-rings on a system be replaced if an extensive retrofit is to be carried out.

NOTE

+ If an O-ring is being changed on one end of a pipe or hose, make sure that the opposite end is not disturbed (twisted or turned). If it is disturbed, it must be replaced. Lubricate all hose and O-ring joints with the oil that is to be used in the system. R1234yf also uses nylon barrier hoses and a gasket-type seal to increase the integrity of the system.

Compressors

The compressor is not generally changed unless it shows signs of wear, e.g. if it is noisy or harsh in operation, or the high-pressure gauge fluctuates badly, indicating damaged valves.

If there is evidence of compressor seal leakage, the seal should be replaced with a compatible seal for that particular refrigerant before the retrofit is begun.

It should be remembered that increased pressure and work loads could be placed on the compressor, and there is a possibility that there may be extra thermal loadings that will exaggerate any pre-existing faults.

Generally an R12 compressor in good operating condition will show excellent durability characteristics on R134a because of the bedding-in process that will have occurred on the mineral oils, as well as the residual mineral oil and chlorine mix that will be coating the internal parts.

The systems compressor should be removed and drained, or the system flushed to remove as much of the old refrigerants oil as possible. A complete system charge of Ester or Roc Oil should then be installed in the system. Note: do not use PAG oils on R12 to R134a retrofits. R134a and R1234yf use the same refrigerant oils and therefore require no changing of oil types. Remember to carry out oil compensation.

Receiver dryers/ accumulators

The material inside the dryer acts as the drying agent and is known as the 'desiccant'. When performing a retrofit it is important to replace the receiver dryer/accumulator with one that is suitable for the new refrigerant type.

It is important to remember to replace the O-rings at the receiver dryer.

High-pressure switches

A retrofitted system will have a different discharge pressure. On a marginal system this difference could be a rise of 5 to 15 per cent, which can be observed on normal operating systems. On well-designed systems a pressure fall could be noted.

A pressure rise can cause the high-pressure switch to activate and disengage the compressor. The high-pressure switch in this case will need to be upgraded to compensate for this pressure rise in a retrofitted system.

A 24 kg/cm^2 (2400 kPa) high-pressure switch is normally replaced by a 28 kg/cm^2 (2800 kPa) switch. A 28 kg/cm^2 (2800 kPa) pressure switch is replaced by a 34 kg/cm^2 (3400 kPa) one.

A medium low-pressure switch does not need to be upgraded. The medium side of the switch activates the condenser fans, and the low-pressure switch disengages the compressor if the suction pressure drops below its specified pressure rating. These can be set at between 50 and 0 kPa, according to the system design. (It is usually 30 to 35 kPa.)

Thermal protector

The thermal protector is placed on the side of some compressors and, if it worked correctly on the original refrigerant, will function satisfactorily on retrofitted refrigerant and will not need to be replaced or adjusted.

Auxiliary fans

It is of no value to identify a condenser air flow fault if it cannot be improved as part of the retrofit. It is important to remember that if, during the initial performance testing, the pressure and therefore the temperature indicates a marginal or poor condensing action, it is advisable to increase the air flow. This change be achieved by using the following **auxiliary fans**:

- adding a condenser fan
- adding an extra condenser fan
- modifying the existing fan to operate whenever the compressor is engaged.

If any or all of the above do not work, replacing the condenser with a more efficient type, e.g. a larger surface area or sub-cooled condenser, is recommended. When fitting an additional condenser fan or modifying the existing wiring circuit of a late-model electronically controlled management system, a surge-or spike-protected relay must be used. These relays are usually polarity sensitive and must be wired exactly according to the manufacturer's instructions.

A viscous fan that engages when the radiator and engine temperatures are high is unlikely to provide adequate air flow for the retrofitted system. An exception to this is when there is an extra fan that has been wired and directly linked to the compressor so that, when the compressor engages, the fan comes on to increase air flow across the condenser.

Refrigerant amount

The amount of refrigerant to be used in a particular system changes when a different refrigerant is used, as the expansion rate of liquid to vapour alters with each refrigerant. LPG has a large expansion therefore a much smaller system charge is used (<200 grams). This can be determined in one of the following ways:

Retrofitting R12 to R134a

This is done using the manufacturer's specifications for an R12 system and then using 80 to 90 per cent of that refrigerant charge as the amount of R134a for that particular system.

Do *not* use less than an 80 per cent charge of refrigerant, as this could result in the system failing on hot or humid days (lack of performance). At worst, the compressor could fail because of excessive superheat levels within the system. Refer to Service Procedure 6 for manufacturers' specifications on system capacities.

Plateau area method

This can be used to determine the correct charge for a specific system (see Figure 6.1).

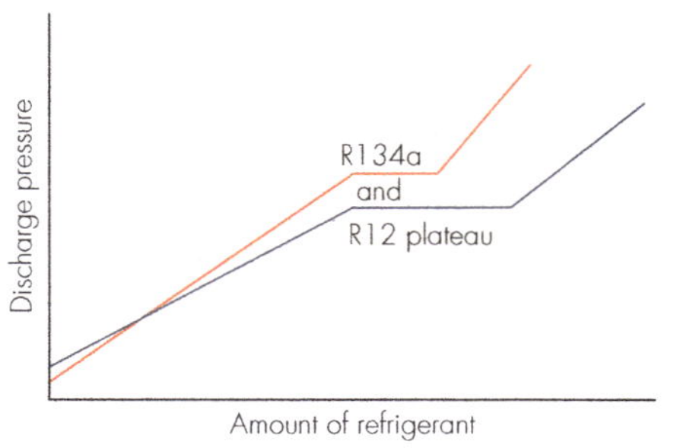

Figure 6.1 Plateau area graph

NOTE

+ The ambient temperature must be 25°C or higher before proceeding with this method. Make sure that you read the complete text before proceeding. This is a basic method for determining the proper refrigerant charge in all refrigerant systems except LPG.

After preparing or servicing the system, charge it with 300 grams of the refrigerant that you intend to use. This should allow the pressure switches to activate the system.

Allow the engine to reach normal operating temperature with:

- the air conditioning on
- an auxiliary fan placed in front of the vehicle, directly in line with the air intake – for **ram air** effect
- the fan speed set to high
- the system set in the recirculate mode
- the engine set at 1500 rpm
- all engine and condenser fans working correctly.

Allow the air-conditioning system to stabilise with this initial charge in it.

At each stage of this procedure, the amount of refrigerant and the discharge pressure must be recorded. Now continue to add refrigerant, 50 grams at a time. Allow the system to stabilise after each addition of a refrigerant charge. Record the amount of refrigerant in the system and the discharge pressure obtained from the system after each addition. If a graph were to be made, it would be similar to Figure 6.1.

When there is no, or very little, pressure rise for an added 50 grams of refrigerant, the system is entering the plateau area. The correct refrigerant charge for a given system is in the middle of the plateau area. The plateau length varies from system to system, but it can be said that the plateau area for R134a is much shorter than that for R12. If another 50 grams of refrigerant causes the pressure to rise again, the system has left the plateau area and is over-charged. Suction line temperature at the evaporator outlet drops to 0°C or less.

If a suitable plateau area cannot be found, the system's condensing capacity must be increased by the use of either a larger condenser or extra air flow.

After the correct refrigerant charge has been determined and the plateau pressure is stable, the amount of refrigerant is correct. If too much refrigerant has been charged into the system, recover, **evacuate** it and recharge to the correct amount.

Discharge pressure method

Use discharge pressure to verify the high-pressure liquid line temperature method. This method can also be used to verify any of the other methods used (see Figure 6.2).

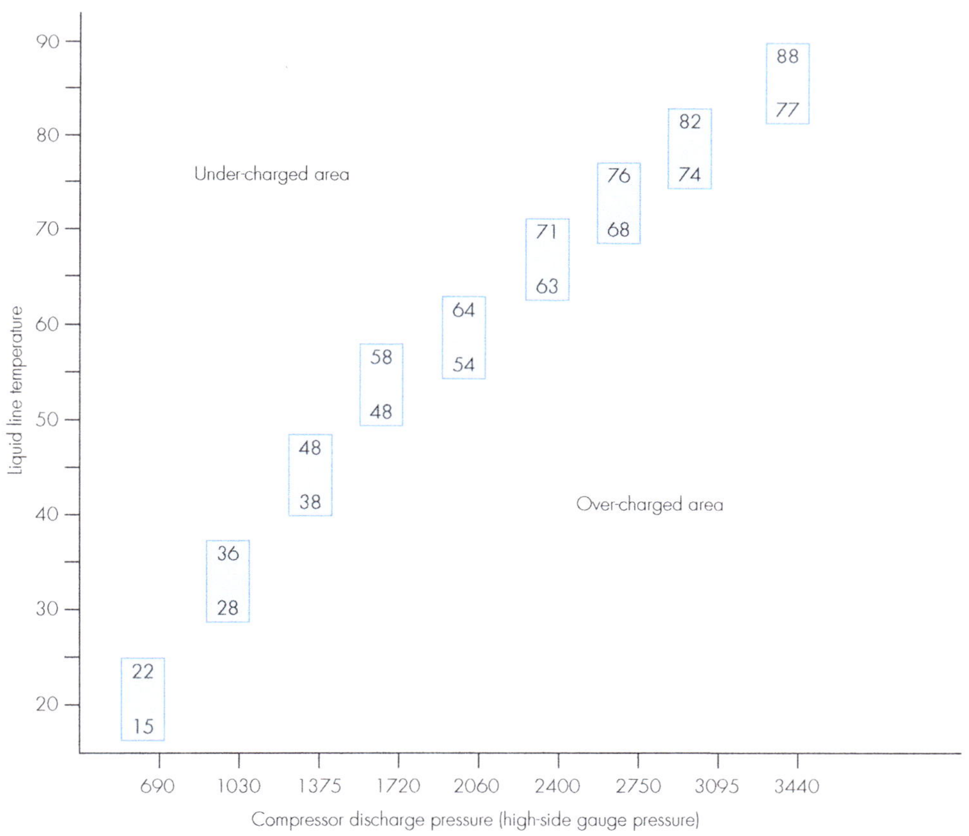

Figure 6.2 Average high-pressure temperature to liquid line temperature of 7–12°C for R134a refrigerant

Retrofitting R1234yf to R134a

This process is against the law, as it will contravene the greenhouse abatement regulations. Therefore this procedure will not be discussed in this book.

Performance testing

A final performance test should now be carried out. It does not need to be as extensive as the previous performance test.

The areas that require close attention are:

- discharge pressure creep (the gauge pressure rising with each cycle) – this indicates inadequate condenser condensing action
- a corresponding drop in cooling performance inside the vehicle.

Tags and labels

The final step in the retrofit process is to label the system to let others know what has been done to it. Labels or tags must be fitted to the vehicle and/or the system.

Place the labels next to the emission decals/tune-up specifications. Tags are usually placed on or near the receiver dryer. The information contained on these labels or tags should include:

- business name
- licence number
- work carried out
- components changed
- refrigerant used
- amount of refrigerant
- lubricant type and amount
- date of service.

Chapter-revision questions

SELF-CHECK QUESTIONS

1. What two tests should be performed on the condenser to determine its efficiency?
2. Before retrofitting a performance test should be completed. What is actually tested?
3. What components may need to be replaced during a retrofit?
4. What is the construction of hoses used in R134a systems?
5. When retrofitting an air-conditioning system which O-rings need replacing?
6. What needs to be done to the compressor when retrofitting?
7. What can be fitted to increase air flow through the condenser after retrofitting?
8. What must be fitted to the vehicle after the retrofitting process has been successfully completed?
9. When refilling the system with refrigerant how much should be added to the system initially before starting the engine?
10. Referring to the previous question, how much should be added at a time after start-up?

CHAPTER 7

Receiver dryers, accumulators, moisture and its removal

Objectives

On completion of this chapter the student will be able to:

- understand the damage that moisture does to the air-conditioning system
- state the amount of moisture that causes this damage
- use the correct servicing methods to remove the maximum amount of moisture from the system
- describe the purposes of a receiver dryer
- identify an accumulator and its position in the system.

This chapter covers some of the essential knowledge and associated skills involved with:

- AURT222670A Servicing air-conditioning systems – evacuation and servicing the receiver dryer/accumulator.

Moisture and its removal

All refrigerants can be considered to be moisture-free. The moisture content of new refrigerant should be less than 10 parts per million (ppm), i.e. 10 litres of water to 1 million litres of refrigerant.

Whenever an air-conditioning system is opened as part of a service in order to replace components, air and moisture are introduced into the system. All refrigerants are hygroscopic (i.e. they absorb moisture). To keep it moisture-free, the system has a receiver dryer or accumulator to remove any moisture from the system. This contains a **molecular sieve**. Molecular sieves can hold up to three times the amount of moisture per unit than older-style desiccants such as **silica gel**.

If more moisture is introduced into the system than the desiccant can absorb, the excess is free to move around the system, and can cause damage to its internal components and block TX vales and orifice tubes rendering the system inoperative.

One drop of water left in the system can cause the moisture content of the refrigerant to be in excess of 40 ppm, or twice the recommended maximum level. Refrigerants react with moisture to form acids.

Heat generated within the system helps to speed up the acid-forming processes. Further damage is caused when oxides form on the internal components, slowing the heat transfer process and restricting the efficiency of the system. They also clog up the receiver dryer.

Not only PAG and **ester oils** are hygroscopic, mineral oil can also add to the moisture problem in the system. Roc Oil 68, on the other hand, is not hygroscopic and is therefore the preferred oil in this case.

If excessive moisture is allowed to stay in the system and is neglected long enough, small holes that allow refrigerant to escape will appear in the condenser and evaporator. Further, the aluminium components, such as pipes and compressor parts, may become unserviceable. The system should be **purged**, evacuated and a new receiver dryer fitted whenever there is evidence of moisture. A system or component flush may also help, as it will restore the cleanliness of the internal areas. Thus, restoring each component to its original cleanliness makes for better heat transfer.

An air-conditioning technician can prevent moisture and **contaminants** from entering the system by:

- installing the receiver dryer last
- capping all parts to keep them air-tight
- never working in wet, humid areas
- keeping refrigerant oil in closed containers
- always evacuating for a minimum of 45 minutes.

It is important that system oil be added whenever any component is replaced. This will compensate for any oil that might have been in the component when it was removed.

Removal of moisture

Moisture is removed from the system by placing the system in a vacuum. In a vacuum, moisture boils and vaporises at a lower temperature. Water boils at 100°C at sea

Figure 7.1 Vacuum pump
Source: http://www.javac.com.au

level, but under pressure the boiling point can be raised. Similarly, it can be lowered by removing pressure (i.e. by placing it in a vacuum).

A good **vacuum pump**, see Figure 7.1, is capable of creating a vacuum of –100.5 kPa, and at this vacuum level water will boil at 4.4°C. As long as the ambient temperature is above this the moisture will be removed as a vapour. Table 7.1 shows the vacuum required to boil off the moisture at certain ambient temperatures.

The recommended system evacuation time is 45 minutes. However, longer evacuation times may be needed if the system has being retrofitted or shows evidence of previous moisture contamination.

When evacuation occurs at higher altitudes, the vacuum pump works at a reduced efficiency, even though moisture boils at a lower temperature. Under these conditions a moisture micron indicator is advisable. It is recommended that the duration of the evacuation be increased in proportion to the altitude above sea level.

Table 7.1 Relationship between vacuum (pressure at sea level) and the boiling point of water

System vacuum at sea level		Boiling point of water
in/Hg	kPa	°C
26.45	-89.6	48.9
27.32	-92.5	43.3
27.99	-94.8	37.8
28.50	-96.5	32.2
28.89	-97.8	26.7
29.18	-98.8	21.1
29.40	-99.6	15.6
29.66	-100.4	10.0
29.71	-100.6	4.4
29.76	-100.8	-1.1
29.82	-101.0	-6.7
29.86	-101.1	-12.2
29.87	-101.2	-15.0

The receiver dryer

The receiver dryer is a very important component of the air-conditioning system, performing three main functions:

- It stores refrigerant and acts as a reservoir.
- It removes and holds moisture from the system.
- It filters out foreign materials.

The receiver dryer is situated either at the condenser outlet or between the condenser and the TX valve, as shown in Figure 7.2.

Three different types of receiver dryer are used in modern automotive air-conditioning systems:

- Standard *upright* units which are shown in Figure 7.3.
- Rare *in-line* types. These are long tubes, rather like a standard receiver dryer placed

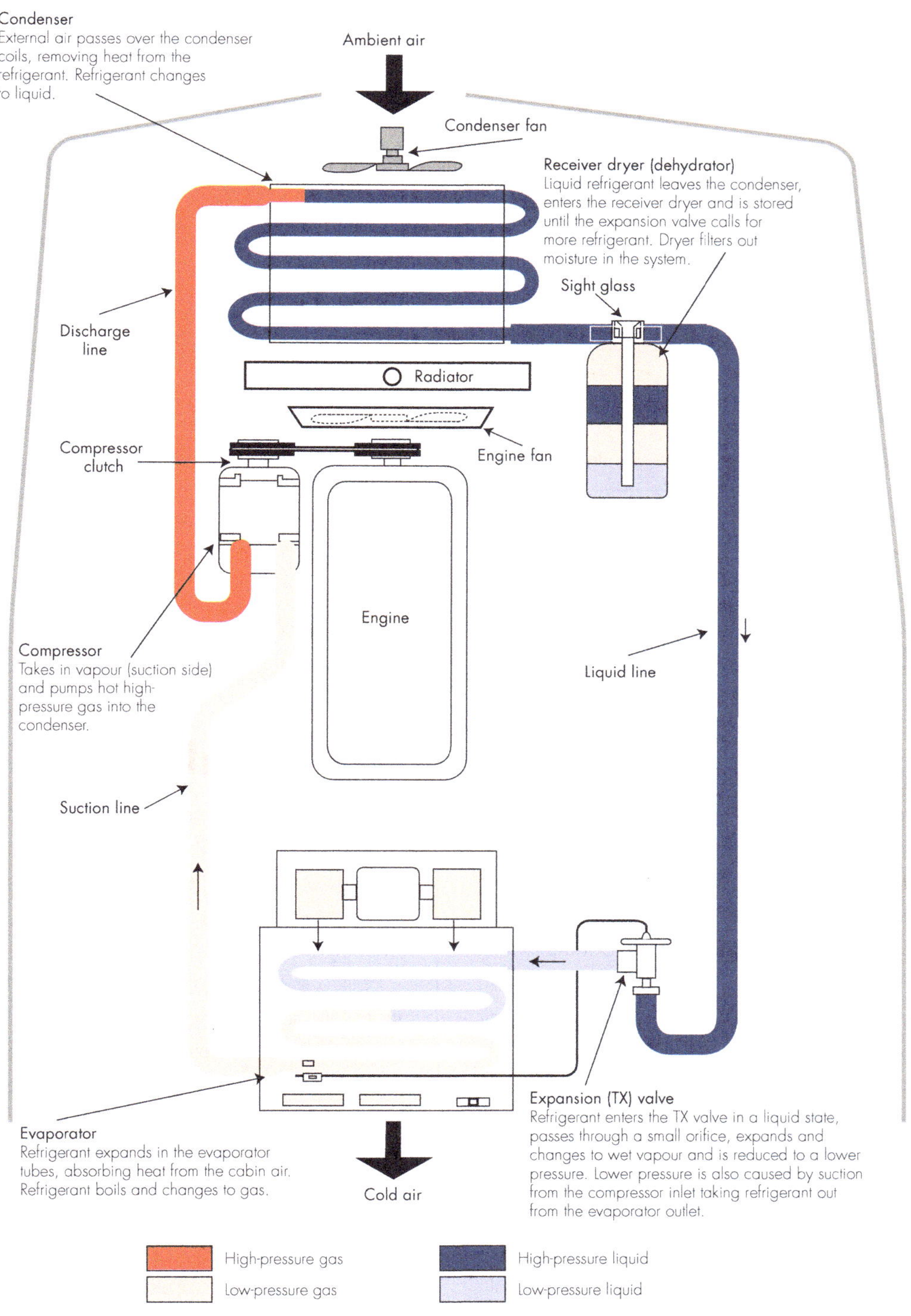

Figure 7.2 Standard TX valve receiver dryer system

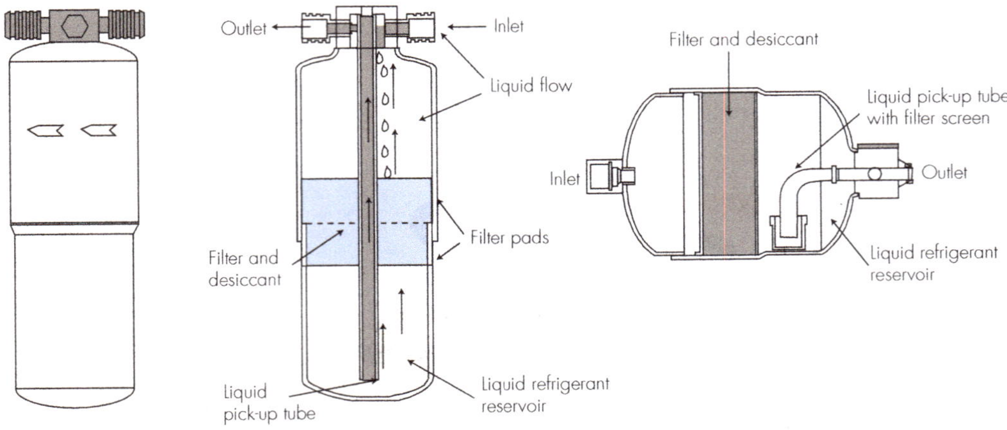

Figure 7.3 Upright and in-line type receiver dryers
Source: Atkins Carlyle Car Parts

on its side and stretched. The inlet is at one end of a unit and the outlet at the other.

› *Accumulators*, which will be discussed later in this chapter.

The receiver part (or container) of a unit stores the system's refrigerant until it is needed by the TX valve, e.g. under conditions of low **head pressure** or low heat load.

As the heat load on the system increases, the unit becomes able to supply the extra refrigerant that it has been holding as needed by the system.

This reservoir action is why it is not necessary to know the exact system refrigerant charge to the nearest gram, and it is good enough to calculate it to the closest 25 grams. This can be seen by looking at the charge plateau chart (Figure 6.2) on page 65.

Across the plateau area even a 50 gram extra charge will make little or no difference to the system's pressure or performance. The desiccant or drying agent in the receiver dryer is to remove moisture, and a **sight glass** is still placed on the outlet side of the unit so that the refrigerant can be seen as it flows into the liquid line.

Desiccant

A desiccant can remove moisture from a vapour, a solid or a liquid. It is placed between two screens or strainers, as shown in Figure 7.4. There are two types of desiccant: silica gel and molecular sieve.

Silica gel is a form of pure silica similar to that found in certain packing containers where moisture needs to be absorbed. Silica gel absorbs moisture when exposed to the atmosphere for even a very short period, and if left open it can become saturated very quickly. Silica gel is no longer used for removing moisture in receiver dryers.

Molecular-sieve is a chemical substance that has cavities, holes or craters on its surfaces, into which moisture is trapped. Moisture molecules are smaller than refrigerant molecules and are therefore easily separated and trapped.

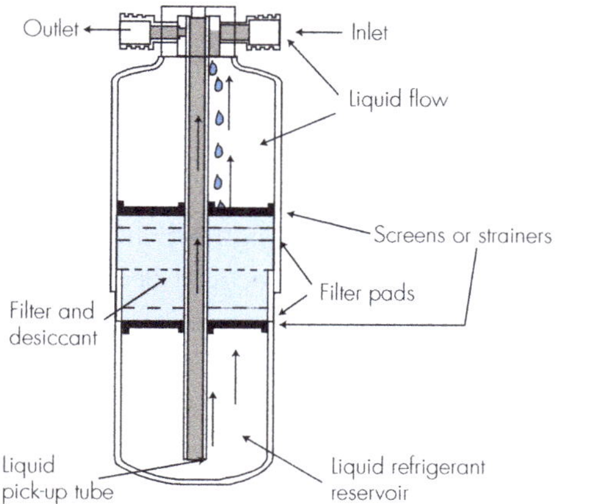

Figure 7.4 Placing of the desiccant in a receiver dryer

This substance is not affected by short exposure to the atmosphere, and can absorb some 300 per cent more moisture by volume than older type systems that used silica gel.

Filters

The receiver dryer usually contains one or two filter pads that the refrigerant has to pass through before it leaves the unit. These filter pads are usually made from a felt substance and they prevent small solids and dust from getting into the desiccant and causing it to decompose as it moves around the system. The outer sides of the filter consist of a strainer made from a wire mesh or gauze material. This holds the filter and the desiccant in place and prevents larger particles from passing through. Both the filter and the desiccant have limited life spans and should be replaced every two or three years. Once the receiver dryer has been saturated with moisture, it cannot be evacuated at normal ambient temperatures. Placing it in an oven at 150°C for one hour allows some moisture removal, but also breaks down the desiccant.

Liquid pick-up pipe

The liquid pick-up pipe ensures that 100 per cent liquid is fed into the high-pressure liquid line and on to the TX valve. The refrigerant that enters the receiver dryer may be a mixture of vapour and liquid. As the liquid is heavier than the vaporous gas, it will fall to the bottom of the unit and so allow the pick-up pipe to supply a constant stream of liquid to the TX valve.

Sight glass

The sight glass is still fitted in the receiver dryer, but should not be used as a diagnostic tool as it can be unreliable.

The sight glass shows the flow of the refrigerant as it passes into the liquid line and on to the TX valve. Sometimes it is a separate unit found elsewhere in the liquid line.

The sight glass should not be used in retrofitted or OEM R134a systems. This is because R134a refrigerant and some of its oils can give a cloudy, unreliable sight glass indication.

Some systems do not even have a sight glass indicator, the charge rates being determined by knowledge of the system's refrigerant capacity, system operating pressures, centre air vent temperatures, component temperatures and compressor cycling times. Using these data together is known as performing a 'performance test'.

How a receiver dryer works

The location of the receiver dryer is important, as its temperature affects its ability to remove and retain moisture. Some systems locate the unit beside the condenser (or even as part of the condenser) in the ambient air flow, while others mount it in the engine bay as far away from any extra heat source as possible. As the temperature of the unit rises, its ability to remove and hold moisture lowers. This means that it is advantageous to keep the receiver dryer as cool as possible.

Moisture that is allowed to roam freely around the system will react with the metals, forming acids that attack them and damage the system.

All moisture must be removed from the system by evacuation in order to prevent any problems. If moisture is left in the system on hotter, more humid days, the desiccant's ability to retain the moisture is lowered, causing it to release droplets into the system. As little as one droplet can cause the TX valve to block or malfunction. The low pressure then drops, the performance (cold air) is reduced, and the low-pressure switch stops the entire system from working.

In the workshop situation it is very hard to identify moisture in the system as the ambient temperatures and heat loads on the system are not as high as they would be under normal working conditions.

The following customer complaints may suggest that moisture is present in the system:

› The system works well in the mornings.
› Around midday the system works fine for 15 minutes or so, after which the performance drops off, the air becomes warmer and the system stops working altogether (low-pressure switch cutting power to the compressor clutch).
› After a while it starts to work well again, before failing once more.
› In the afternoon the system once again works well.

Diagnosis

Moisture turns to ice as it moves around the system, blocking the TX valve.

If a temperature difference is found across the receiver dryer, this is a sign that there is an internal restriction and the receiver dryer must be replaced. Place a temperature probe at the unit's inlet and record its temperature, then measure the temperature at the receiver's outlet. There should be no more than 1–2°C difference.

Installation and servicing

As already mentioned, the receiver dryer is either mounted near the condenser outlet or in the engine bay away from any heat source. The various types and uses of hose fitting include:

› flare with nuts
› O-rings
› barbs and clamps
› a combination of the three.

Receiver dryers are directionally sensitive. The inlet is clearly marked, and must be connected to the pipe that comes from the condenser. Reversing the direction of the receiver dryer could lead to vapour being transferred to the TX valve and a subsequent drop in cooling efficiency.

The upright type receiver dryer must not be mounted at an angle greater than 15°C.

When servicing the system, the receiver dryer must be the last component to be connected, as moisture can start to enter the desiccant and shorten its expected life. Once the receiver dryer has been connected, evacuation should be commenced immediately and continued for at least 30 minutes. If the system being serviced is retrofitted, or if it has previously had a moisture problem, it should be evacuated for up to one hour.

NOTE

+ All connections must be coated with the same oil that is being used in the system. This will avoid damaging the joints' surfaces and possibly causing refrigerant leakage.

Accumulators

Accumulators are similar to receiver dryers in that they have filters, desiccants and screens (see Figure 7.5). The main differences are:

› Accumulators are positioned in the return (low-pressure) line between the evaporator and the compressor.
› Their purpose is to accumulate any liquid and make sure that it has vaporised before it returns to the compressor. This is the opposite of a receiver dryer's action.
› Accumulators allow liquids to change to vapours.

It can be seen that the refrigerant enters the accumulator and will drop to the bottom of the unit if any liquid is present. The large surface area allows the liquid to boil off (vaporise) and then rise to the outlet port and return to the compressor. There is a strainer and oil bleed hole at the bottom of the accumulator, allowing a small amount of oil to return to the compressor if a blockage occurs within the accumulator. This maintains compressor lubrication at all times.

Both the receiver dryer and the accumulator are not serviceable – if they are found to be faulty, they must be replaced. Common faults of receiver dryers and accumulators include:

› leaks at joints and welded sections
› blockages caused by moisture, particles, contamination and incorrect positioning
› fitting in the reverse direction
› deterioration due to age.

Receiver dryers should be replaced every two to three years and accumulators every three to four years.

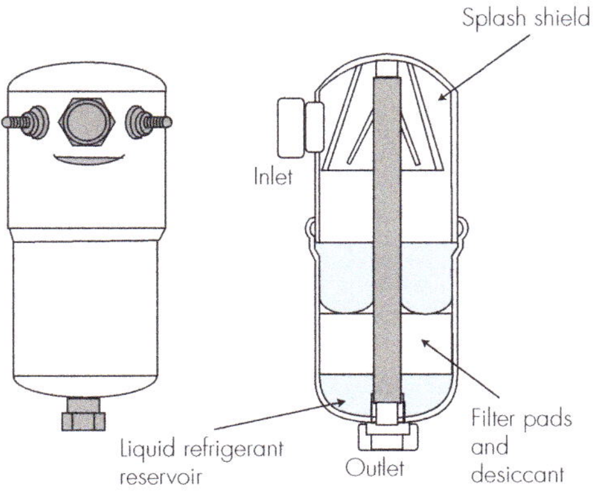

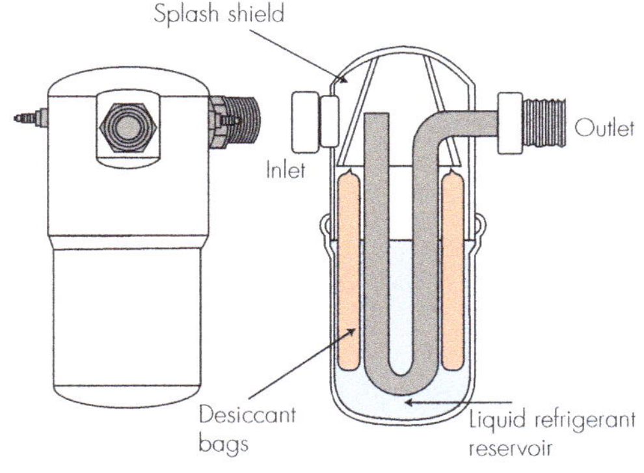

Figure 7.5 Typical accumulators
Source: Atkins Carlyle Car Parts

Chapter-revision questions

SELF-CHECK QUESTIONS

1. What is the recommended moisture content for new refrigerant?
2. What does the term 'hygroscopic' mean?
3. What components in an air-conditioning system are designed to remove and store excess moisture?
4. How is moisture removed from an air-conditioning system?
5. How can moisture be prevented from entering an air-conditioning system?
6. How can moisture affect air-conditioning systems?
7. Why should oil be added to the system when replacing components?
8. What is the recommended evacuation time?
9. When should evacuation times be increased?
10. What would the boiling point of moisture be if a vacuum pump only went down to −99.6 kPa?
11. What are the main functions of a receiver dryer/accumulator?
12. What is the function of the oil bleed hole used in an accumulator?
13. What are three ways of connecting hoses to receiver dryers?
14. What are two types of desiccant used in receiver dryers?
15. What are some common faults found with receiver dryers/accumulators?

CHAPTER 8

Service valves and manifold and gauge sets

Objectives

On completion of this chapter the student will be able to:

- recognise the different types of service valves and their uses
- describe the differences between high- and low-pressure gauges
- calibrate the gauge sets
- service the pressure hoses to prevent refrigerant leakage.

This chapter covers some of the essential knowledge and associated skills involved with:

- AURT222670A Service air-conditioning systems.

Service valves

Service valves are devices that allow the service technician to use mechanical means to open the air-conditioning system. This makes it possible to connect a **manifold** and **gauge set** to the system to perform diagnosis testing, as well as removing and replacing refrigerants.

Most air-conditioning systems have two service valves, one on the low-pressure side of the system between the evaporator and the compressor, and the other on the high-pressure side between the compressor and the TX valve. There are some older systems, however, that use a second service valve on the low-pressure side of the system to measure the pressure drop across the suction pressure regulator, but this is used only for diagnosis purposes.

Types of service valves seen on modern air-conditioned vehicles:

- **Quick-release connectors** – these are used for R134a and R134a-retrofitted systems.
- New refrigerant quick-release connectors – similar to R134a quick-release connectors, but with a different outer connector profile.

Schrader valves

Figure 8.1 shows the **Schrader valve**, which was for many years the most common style of service valve. However, with the introduction of new refrigerants such as R134a, the quick-release connector fitting has become the most common service valve – though it still uses a Schrader valve inside the fitting.

A Schrader valve is very similar to a tyre valve (although they should not be interchanged). It has two positions: open and closed. When a **manifold hose** is connected, a pin in the hose fitting depresses the valve, which then opens to allow the refrigerant to flow freely to the manifold and gauges.

NOTE

+ All hose fittings must be leak proof. They must not allow excessive refrigerant to escape while being disconnected.

The *EPA Ozone Protection Act* states that the service fitting cap *must* be fitted so as to prevent refrigerant leaks. It is recommended that the Schrader valve cores be replaced whenever the system is serviced, as they can be a common cause of refrigerant leakage.

Quick-release connector service valves

Quick-release connectors are similar in principle to the popular garden hose clip-on connectors,

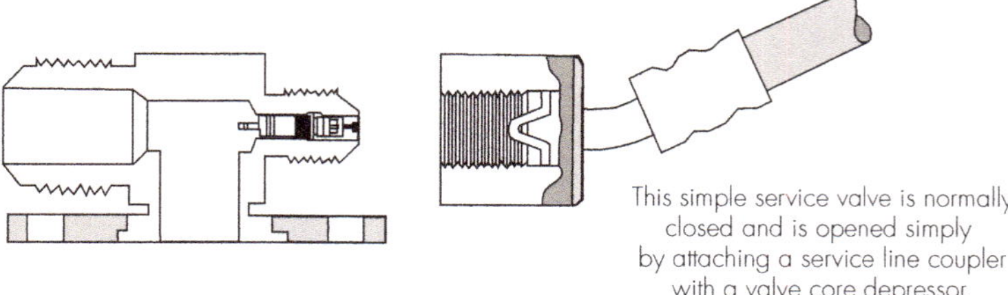

Figure 8.1 R12 Schrader type service valve
Source: Atkins Carlyle Car Parts

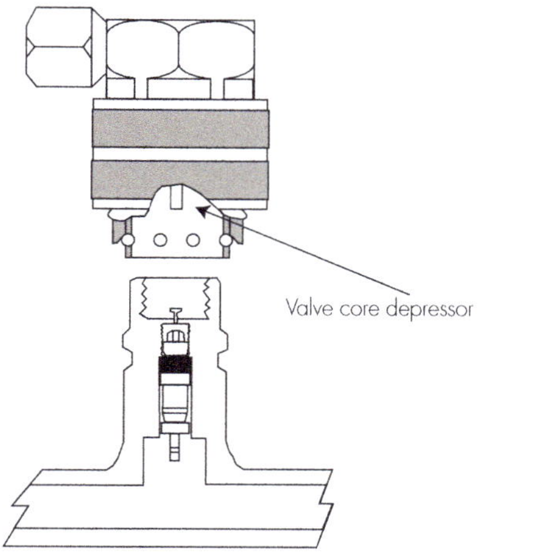

Figure 8.2 R134a quick-release coupler
Source: Atkins Carlyle Car Parts

and are designed for use with HFC R134a systems (see Figure 8.2).

The valve itself is still a Schrader design – there are at least four different Schrader valve designs in use (see Figure 8.3). The valve housing is designed so that only the correct connector can fit. The high-and low-side connectors are different sizes in order to avoid incorrect connection to gauges (see Figure 8.4).

Service Procedure 1 on page 186 shows how to fit the quick-release connectors that connect the manifold and gauge set.

New refrigerant quick-release connector service valves

All new refrigerants, such as HFO1234yf, will have service connectors that have a slightly different profile and diameter so as to prevent cross-contamination of refrigerants.

Summary of service valves

Service valves are prone to leak and it is important to make sure that the dust caps are refitted to prevent any entry of dirt or moisture and to help prevent refrigerant leakage.

There is no set place on an air-conditioning system for positioning the service valves, but the law now states that they must be in an accessible location.

Manifold and gauge sets

The most important tool that an air-conditioning technician will use is the manifold and gauge set. This provides the technician with the means to accurately measure system pressures, and

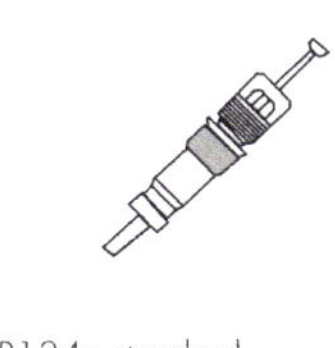

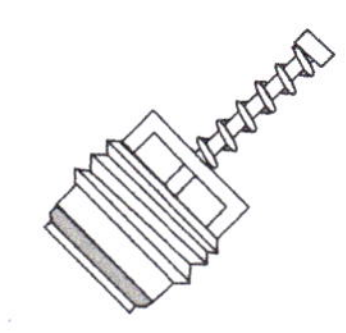

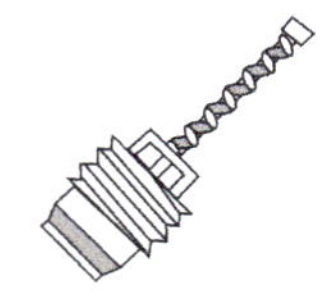

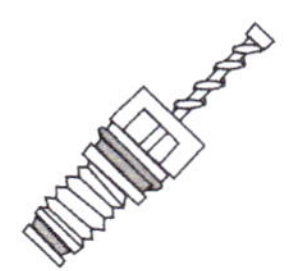

Figure 8.3 Four different types of Schrader valve
Source: Atkins Carlyle Car Parts

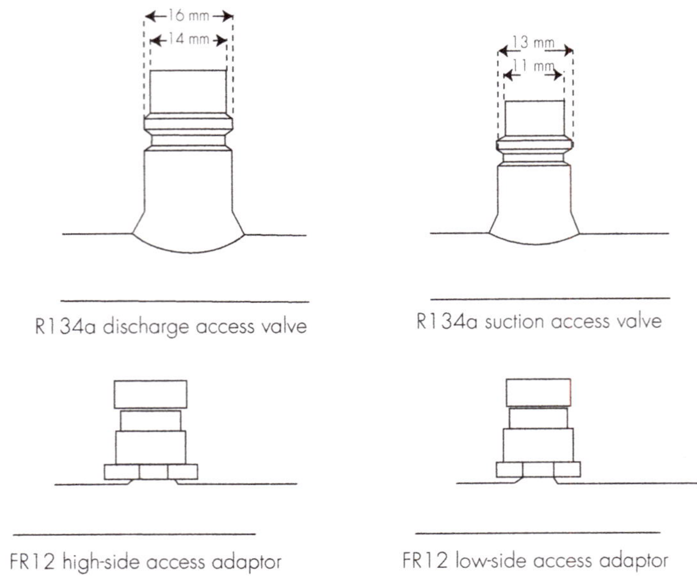

Figure 8.4 Blended refrigerant quick-release connector service valves
Source: Atkins Carlyle Car Parts

therefore assess the system's performance. The manifold and hand valves allow the system to be purged of refrigerant, evacuated of air and moisture, and recharged with new refrigerant.

Most modern gauge sets use two gauges, but some air-conditioning systems that use a pressure control regulator for the evaporator may require a second **low-pressure gauge**.

The normal manifold and gauge set has two gauges (refer to Figure 8.5):

- low-pressure gauge (combination pressure and vacuum)
- **high-pressure gauge** (pressure only).

Each of the gauges is connected to the system via the manifold, hand taps and hoses.

Each hand tap, when opened, allows the refrigerant to flow out of the system into the manifold and into the centre yellow service hose.

The hand valves and yellow hose are used to purge, evacuate and charge the system.

The hand valves *must* remain closed at all times so that it is possible to read the system pressures. They should only be opened to service the system.

The low-side gauge

The gauge used on the low-pressure side of the system is called a **compound gauge** (refer to Figure 8.5). It gives both pressure and vacuum readings. It connects to the blue hose through the manifold.

The metric scale starts at −100 kPa and has a maximum pressure accuracy reading of 800 kPa (−1 to 8 bar), but the gauge can read up to 2400 kPa or 24 bar.

The imperial scale reads vacuum pressures between 30 and 0 inches of mercury (in/Hg), and pressure from 0 to 300 psi.

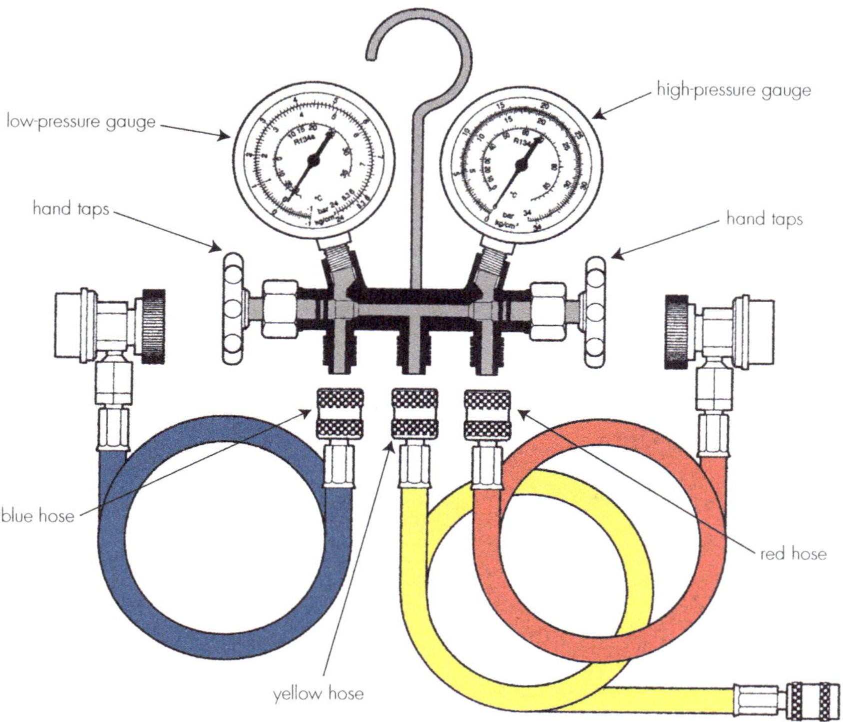

Figure 8.5 A modern two-gauge set
Source: Atkins Carlyle Car Parts

The high-side gauge

The **high-side gauge** shows the pressure in the high-pressure side of the system, and is connected to the red hose through the manifold. It does not read vacuum and is therefore not a compound gauge. It cannot be damaged by applying a vacuum to it. The metric scale reads from 0 kPa to a maximum of 3500 kPa. The imperial scale reads from 0 to 500 psi.

Gauge calibration

Gauges are equipped with **gauge calibration** adjustment screws to allow for changes in atmospheric pressure and spring tension due to having refrigerant pressure on it most of the time. Most good quality gauges are reasonably accurate (to about 2 per cent of the total scale reading) and are vibration-dampened. The gauge needle needs to be adjusted so that the needle rests on zero when there is no pressure on the gauge.

NOTE

- Adjusting the needle to zero only adjusts the gauge at zero *not* total scale reading. If there is a requirement for a good accurate reading, a class 1 gauge should be used, calibrated with a dead weight tester.

To calibrate a gauge, simply remove the glass or plastic cover and retaining ring (or 'bezel'). Make sure that there is no pressure in the hose

Figure 8.6 Adjusting the calibration on a gauge

before making any adjustments. Using a small screwdriver, turn the adjusting screw either way (see Figure 8.6) until the needle lines up with zero. Make sure that you look directly over the needle to avoid the parallax error. Do not try to force the adjusting screw, as it is easily damaged.

Hoses

The **charging hoses** are constructed to withstand working pressures in excess of 3500 kPa. Some hoses have reached pressures as high as 13 800 kPa before bursting.

A standard colour code is used when connecting hoses. Blue is used on the low-side pressure, red on the high-side pressure, and yellow is for the service hose. The colour-coded hoses are for ease of connecting to the system and other equipment. Under current ozone protection legislation all refrigerant hoses must have an isolating valve to hold the refrigerant in the hose when it is disconnected from the system.

Hose fittings can be easily damaged and therefore create leakage points during evacuation and charging procedures. They should be serviced or replaced periodically. Hose connectors and hand taps should only be hand-tightened.

NOTE

+ Although these hoses are used by default for evacuation, they are not designed for this task, so obtain a first-class evacuation-use vacuum hose (very expensive) or a purpose-made refrigeration copper tube, which can be remade as it gets worn or damaged.

Chapter-revision questions

EXERCISES

1 Check a manifold and gauge set to ensure that they are reading zero, and then practise adjusting them (refer to the section on gauge calibration).
2 Check the charging hose fittings to make sure that they are serviceable and not leaking.
3 Connect a set of gauges to an air-conditioning system as described in Service Procedure 1 on page 186.
4 Carry out a performance test in accordance with Service Procedure 2 on page 189, taking particular note of the following:
   - the low-pressure reading
   - the high-pressure reading
   - the ambient temperature
   - the centre outlet temperature.

SELF-CHECK QUESTIONS

1 What is the highest reading that can be taken on the high pressure gauge?
2 Aside from the maximum scale on low- and high-side gauges what is the main difference between the two gauges?
3 When taking pressure readings how far should hand valves be opened?
4 Why do gauges need calibration?
5 What are the main functions of a manifold and gauge set?
6 Where on the air-conditioning system is the low-pressure service point found?
7 Where is the high-pressure service point found?
8 What is a compound gauge?
9 What is the procedure to calibrate a gauge?
10 What is the function of the blue, yellow and red charging hoses?

CHAPTER 9

Leak detection

Objectives

On completion of this chapter the student will be able to:

- distinguish between the various types of leak detectors: electronic, nitrogen gas and refrigerant dye
- state the difference between the types of leak detectors and their accuracy
- describe the advantages of dyes and nitrogen gas in finding small leaks.

This chapter covers some of the essential knowledge and associated skills involved with:

- AURT222670A Service air-conditioning systems – leak detection

The most common complaint that air-conditioning technicians receive from customers is a lack of cooling. This is almost always due to a lack of refrigerant within the system – in other words, a system refrigerant leak.

There are several ways of detecting refrigerant leaks, and we will describe them here.

Electronic leak detectors

In Australia, the Ozone Protection Act requires that air-conditioning technicians have at least an **electronic leak detector** as part of their leak detection equipment (see Figure 9.1). Modern electronic leak detectors can find leaks as small as 0.5 grams a year. The Australian laws can be found at http://www.arctick.org.

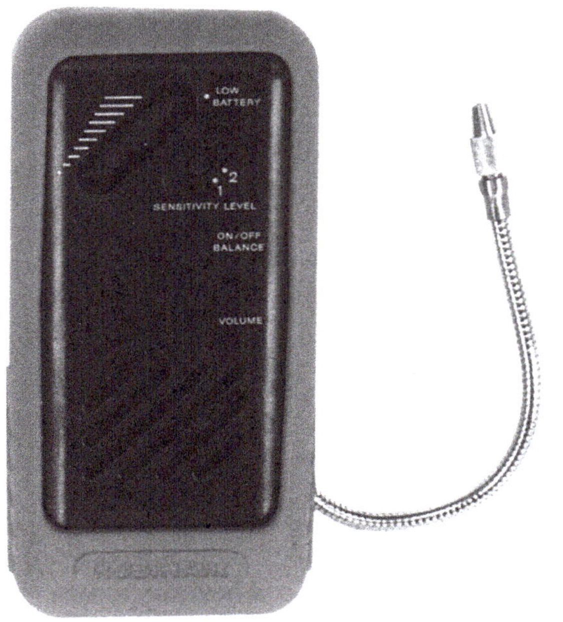

Figure 9.1 An electronic leak detector
Source: Atkins Carlyle Car Parts

In New Zealand the release of a controlled substance (including CFCs) into the atmosphere is prohibited and the only way to detect a release into the atmosphere is to use leak detection techniques. The New Zealand legislation can be found at http://www.mfe.govt.nz/issues/ozone/depleting-substances/cfc-mgmt-strategy.html

Electronic leak detectors work by one of two methods:

- On the *negative corona discharge principle* of applying a DC voltage of approximately 2000 volts across an electrode. When refrigerant is present, the current flow across the electrode is decreased and the electronics in the detector increase the tick rate. If the leak is severe enough, a siren will go off. These types of leak detectors can find a leak as small as less than 1 gram per year.
- *Electrochemical heated diode sensors* incorporate refrigerant-selective technology. These are more accurate and reliable than their predecessors and are able to trace refrigerant leaks as small as 0.5 grams per year.

Whichever electronic leak detector is used, the service and use is similar. It is essential to keep the tip of the sensing detector clean and free of grease. Breathing into the sensing tip can severely shorten its life expectancy. Moisture, water or oil, will destroy the sensing tip.

The service life of sensing tips is between six and nine months depending on use. They must be replaced regularly if you are to be certain of the unit's accuracy. Always carry out a prove-test before use check and prove that the electronic leak detector actually works.

Using electronic leak detectors

- The system must have a head pressure of 350 kPa.

The leak detector should be switched on and given time to warm up, which could take between one and two minutes. The air being sensed around the unit will register as normal. However, when the detector is moved closer to a leak it will sense a difference in the resistance across its electrode or diode and sound an alarm. Do not turn the detector on while it is situated close to a leak as it will then register as normal and it is possible that the sensing tip could be damaged.

NOTE

+ Carpets and vinyl give off gases of their own that could cause a false siren to sound when there is, in fact, no leak. To avoid this, turn the detector on inside the passenger compartment away from any refrigerant component so that the detector will sense the carpet or vinyl gases as normal.

The sensitivity of the unit should be adjusted so that the leak detector ticks only two or three times per second. As a rule, the sensitivity of the detector should be set to minimum and only increased for very small and difficult leaks. It should be returned to minimum when the leaks have been found.

Remember that refrigerants are heavier than air. There is therefore a need to check under each pipe and connection.

The detector should be moved around the air-conditioning system, while being very careful not to touch or rub the surface, but still remaining within 5 mm of the surface. A good leak detector has a rubber or plastic cover fitted over the sensing tip to protect it from any dirt or oil entering into the sensor tip. You should move it around at the rate of approximately 25–50 mm/sec. Always start from one point in the system and trace the complete system so as not to miss any single point. Any signs of oil leakage should be carefully checked as they usually indicate refrigerant loss. Move carefully and methodically around the system, taking special care of all pipes, seals, connections, welded or brazed joints and **service ports** or switches.

The hardest leaks to find are often evaporator leaks, as they are usually inaccessible. Inserting the probe into the drain tube may detect larger evaporator leaks. It may be necessary to remove the **resistor** block for the blower fan (if this is possible) and to insert the probe into the evaporator housing in order to try to detect any refrigerant leaks from that area. If this is not possible, the probe may need to be inserted into the footwell vents to try to detect the presence of any refrigerant. The low-side pressure may need to be increased in order to make it easier to find a leak. This can be done by allowing the system to operate for one or two minutes. Turn the unit off. Opening both manifold hand valves allows higher pressures into the low side so that any small leaks can be found by probing the lower footwell vents, drain tube or fan resister areas.

Ultraviolet dye leak detection

The most efficient method of leak detection is the **ultraviolet (UV) dye** leak detection method (see Figure 9.2). The dye is introduced into the system by a pre-measured tube, or by using an

TO-25675
240 volt model

Injector gun
Shown with quick coupler
Can be supplied with nozzle
for easy injection directly into suction line

TO-25667
Super 400 watt model
12 volt and 240 volt operation

Glo-Leak ultraviolet leak detector

Figure 9.2 Ultraviolet dye leak detector
Source: Atkins Carlyle Car Parts

infuser that is placed as part of the low-pressure service line. The dye mixes readily with the refrigerant oil and, therefore, must be compatible with that oil. The system is run for between 10 minutes and 24 hours to allow the dye to escape at the point of the leak, after which a UV lamp is shone on the areas that are suspected of leaking refrigerant. The UV lamp shows up any leaks as either a yellow or green stain, depending on which product is being used. This stain is invisible under normal light.

After the leak has been repaired, the stain is removed by cleaning with solvent and the system rechecked.

Evaporator leaks can be detected by checking for signs of stain around drain tubes and evaporator housings under the dashboard. This method of leak detection allows the air-conditioning technician to add the dye to the system at a relatively small cost, during any service of the system. The lamp can then be used to check for any leaks after service and, as the dye stays permanently in the system, it becomes a simple task to check for any subsequent leaks that may develop.

If using this method, ensure that a label is affixed to the system to prevent subsequent charging of dye.

Leak detection using nitrogen gas

One of the best methods of finding small leaks is by using nitrogen gas (see Figure 9.3); it is cheap, relatively safe for the environment and effective.

Any remaining gas should be recovered from the system that is suspected to be leaking and a quick evacuation performed to place the system

Figure 9.3 Nitrogen test kit
Source: Reece

Figure 9.4 Cal Blue leak detector fluid
Source: Image courtesy of Nu-Calgon

under a vacuum. The nitrogen manifold and hose should be connected to the system and nitrogen added until the pressure reads between 1600–1800 kPa, do *not* start the engine, the system pressure will equalise on both the high and low sides of the system.

A solution such as Cal-blue (see Figure 9.4) and a paintbrush wiped around joints is a good method of confirming small leaks by displaying bubbles, and to check all joints and connections around the entire air-conditioning system. Larger leaks may be heard with a low hissing sound and confirmed with the solution. Leaks are easier to find because of the high pressure and the fact that the engine fan is not blowing refrigerant gas away.

Once a leak is found and repaired it is important to check the system for any further leaks and repair and check them. Remember: if a system is leaking it is highly likely that there is more than one leak.

NOTE

+ Do not use soap and water, as soap contains glycerine, which can mask a leak.

Chapter-revision questions

PRACTICAL EXERCISE

Using the steps outlined in Service Procedure 6 on p. 205, leak test an air-conditioning system making use of one of the detectors described in this chapter.

SELF-CHECK QUESTIONS

1 What is the most common customer complaint regarding air-conditioning operation?
2 What are the most common types of leak detector?
3 When using an electronic leak detector, which component must be replaced every 6 to 9 months?
4 Why are leaks from evaporators hard to find?
5 When checking leaks using an electronic leak detector where, on pipes and hoses, should you check?
6 How can evaporator leaks be found?
7 What is a major advantage of using UV dye?
8 How much refrigerant should be added to the system when checking for leaks using nitrogen gas?
9 How much nitrogen should be added when leak detecting?
10 Once a leak is found what should be done after repairing the leak?

CHAPTER 10

Compressors

Objectives

On completion of this chapter the student will be able to:

- describe the function of the compressor
- describe the various types of design available
- recognise the types of belts and pulley drives from the different vehicle manufacturers
- understand the operation of the variable displacement compressor.

This chapter covers some of the essential knowledge and associated skills involved with:

- AURT222670A Service air-conditioning systems – compressor operation
- AURT366108A Carry out diagnostic procedures.

This chapter covers the function of the compressor, and demonstrates the many types found in modern vehicles.

The compressor's functions

In simplified terms, the compressor in an automotive air-conditioning system is a vapour refrigerant pump that moves the refrigerant around it.

There are two main functions that a compressor must be capable of achieving. First, it must create a low-pressure area. Working with the expansion valve, the compressor inlet draws heat-laden vapour refrigerant from the evaporator. Second, the refrigerant must be compressed, which results in the low-pressure vapour becoming a high-pressure vapour; i.e. the pressure and temperature of the refrigerant rises and becomes superheated.

If either of these functions fails or becomes inefficient, the air-conditioning system becomes inefficient or inoperative.

Air-conditioning compressors are of three basic designs:

- reciprocating piston
- scroll
- vane rotary.

Although the size, design and number of cylinders may vary, all compressors fit into one of these categories.

Reciprocating piston compressor

The term 'reciprocating' literally means 'up and down', but it can also mean 'back and forth'. Both terms are applicable to types of air-conditioning compressor. 'Crankshaft drive' compressors use up and down piston movement, while 'axial plate drive', 'swash' or 'wobble plate' compressors move back and forth.

Crankshaft drive

This type of compressor is similar in operation to an automotive engine except that the compressor crankshaft drives the piston and it has **reed valves** similar to a two-stroke engine. The crankshaft in the compressor is driven by the engine via pulleys and belts, and can be connected and disconnected from the engine by an **electromagnetic clutch**. This feature is common to all compressors, regardless of type of drive.

Axial plate drive

The axial plate drive is the other compressor design. This operates by using a compressor crankshaft with a plate fixed at an angle of approximately 65° or one that can vary the angle plate either by mechanical or electronic controls. As the crankshaft turns, the profile of the axial plate moves back and forth. The pistons are attached to the plate, resulting in a reciprocating, or back and forth, motion of the pistons.

Operation of reciprocating piston compressors

Crankshaft drives commonly use two cylinders, but axial plate drive compressors may use four, five, seven or ten cylinders. Some designs use five double-ended pistons, which makes a ten-cylinder compressor.

The function of a compressor is to create a low-pressure area at the compressor inlet and to compress the refrigerant vapour into a high-pressure vapour. To do this, each cylinder needs a piston, sealing rings and a suction- and discharge-valve. These valves are usually reed valves that open and close with changes in the cylinder pressure.

As the piston begins to move down, the pressure in the cylinder begins to drop to a

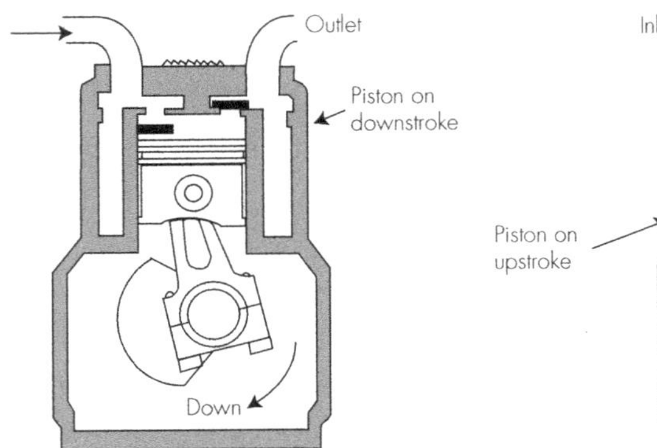

Figure 10.1 The crankshaft drive compressor
Source: Atkins Carlyle Car Parts

Figure 10.2 An axial plate drive compressor

partial vacuum. Because the pressure below the inlet valve is less than the pressure above it, the spring tension of the reed valve is overcome and the valve opened. Heat-laden low-pressure refrigerant vapour fills the cylinder. The discharge valve is kept closed because of the low pressure in the cylinder and the higher pressure above the valve (see Figures 10.1 and 10.2).

When the piston is at the bottom of its stroke and starting to rise, the cylinder pressure also begins to rise, thus closing the inlet valve. Both valves are now closed. When the pressure in the cylinder becomes greater than the spring tension in the discharge valve, the valve opens and high-pressure refrigerant vapour is discharged from the compressor and sent on to the condenser.

It is important that both the sealing ring around the piston and the reed valve spring are in good condition, as any leakage would lead to poor operation. This will become obvious during performance testing of the system by:

› a lack of system performance
› pressure gauge fluctuation discharge and/ or suction gauges fluctuating, which indicates that that reed valve is not sealing.

The compressor is lubricated by refrigerant oil, some of which is stored at the bottom of the compressor, while the rest circulates around the system with the refrigerant.

Scroll compressor

This type of compressor is based on the scroll compression concept (see Figure 10.3). The Sanden company has developed this scroll principle into compressors that give a high rate of efficiency, with a lower rate of noise and vibration than other compressors have.

The design consists of two scrolls – one fixed and the other movable – that mesh with each other. The movable scroll does not rotate but moves in an eccentric action so that the space between the scrolls changes as the movable scroll orbits. It gradually moves towards the centre of the scroll, compressing the refrigerant as it goes. The pressure then opens a discharge valve in the centre of the fixed scroll. This design does not require the use of an inlet valve. This type of compressor is very susceptible to liquid or over-oiling damage.

Vane rotary compressor

The vane rotary compressor does not use sealing rings like a reciprocating piston or scroll compressor. Instead, it uses vanes that are fitted into slots in a rotor, sealed against the housing by both centrifugal force and lubricating oil. The rotor spins in a housing that is off-centre to the rotor, thus making the pumping chamber reduce in size as the rotor turns, and so compressing the refrigerant.

As in the scroll-type compressor, the vane design does not use an inlet valve – only a discharge valve. The discharge valve also serves as a check valve that prevents high-pressure refrigerant from entering the compressor when the compressor is cycled off (see Figure 10.4).

This type of compressor relies on a good supply of refrigerant oil to seal the vanes against the housing, as severe damage will occur if the refrigerant oil is lost as the result of refrigerant leaks, or not topping up during filter changes. It is therefore important that a low-pressure cut-out switch is used with this unit.

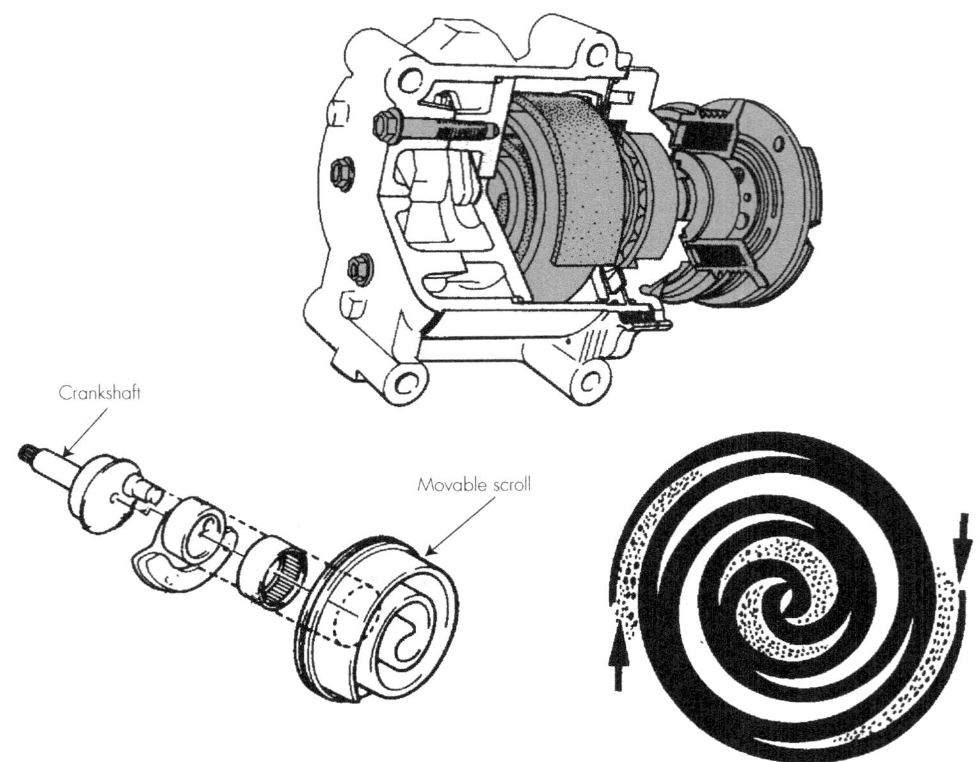

Figure 10.3 The scroll compressor and the scroll principle
Source: Sanden International (Australia) Pty Ltd

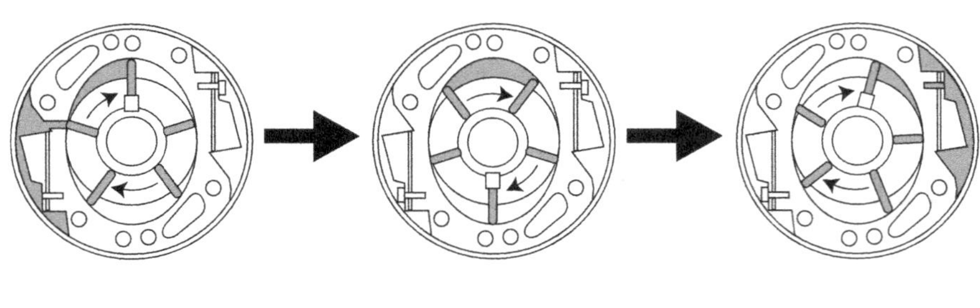

Figure 10.4 The Seiko-Seiki vane rotary compressor
Source: Atkins Carlyle Car Parts

Variable displacement compressors

Compressors have been developed that are continuously driven by the engine. This ensures that cabin temperature and system performance are precisely controlled by varying the **displacement** of the compressor rather than by cycling the compressor on and off.

In the earlier axial plate drive compressor the angle of the plate was fixed. Today, the piston stroke can be varied simply by changing the angle of this plate.

The angle of the axial plate is changed by a mechanical or electronic control valve located in the compressor housing. The mechanical control valve senses suction pressure/temperature because:

- the suction pressure is above the control pressure point when the compressor capacity demand is high, and the compressor is therefore at the maximum displacement position
- the suction pressure reaches the control point when the compressor capacity demand is low, and the valve bleeds discharge vapour refrigerant into the crankcase below the pistons, which closes off the passage to the suction plenum chamber. This pressure differential creates a total force on the pistons that results in movement around the axial plate pivot, a reduction of the plate angle and a reduction in the compressor pumping capacity. The pumping stroke can be varied from 1 per cent to 100 per cent capacity while continuously allowing for oil and refrigerant circulation.

Figure 10.5 demonstrates the action of the **variable displacement compressor**.

The temperature is controlled by varying the capacity of the compressor, and not by cycling the compressor on and off. This action provides a more uniform method of temperature control, while reducing fuel use and eliminating some of the noise problems associated with a cycling clutch compressor system. The suction pressure is higher than a cycling clutch system to eliminate evaporator freezing.

Compressor stroke control

As already mentioned, there are two types of stroke control valves in current use: mechanical and electronic.

Mechanical control valve

The advantage of the mechanical control valve compressor is that it can be used with any simple air-conditioning system, whether it uses a manual control or an automatic temperature control (**ATC**) system. The compressor operation does not need to be controlled by the ECU (electronic control unit). The disadvantage is that it requires a full charge to operate the valve mechanism correctly.

Electronic control valve

The stroke length, and therefore the capacity, of the electronic control valve compressor is controlled by the vehicle's body control module in conjunction with the engine management unit in a similar manner to that of the mechanical control valve compressor, except that the position and size of the valve orifice is controlled electronically by a pulse-width modulated solenoid.

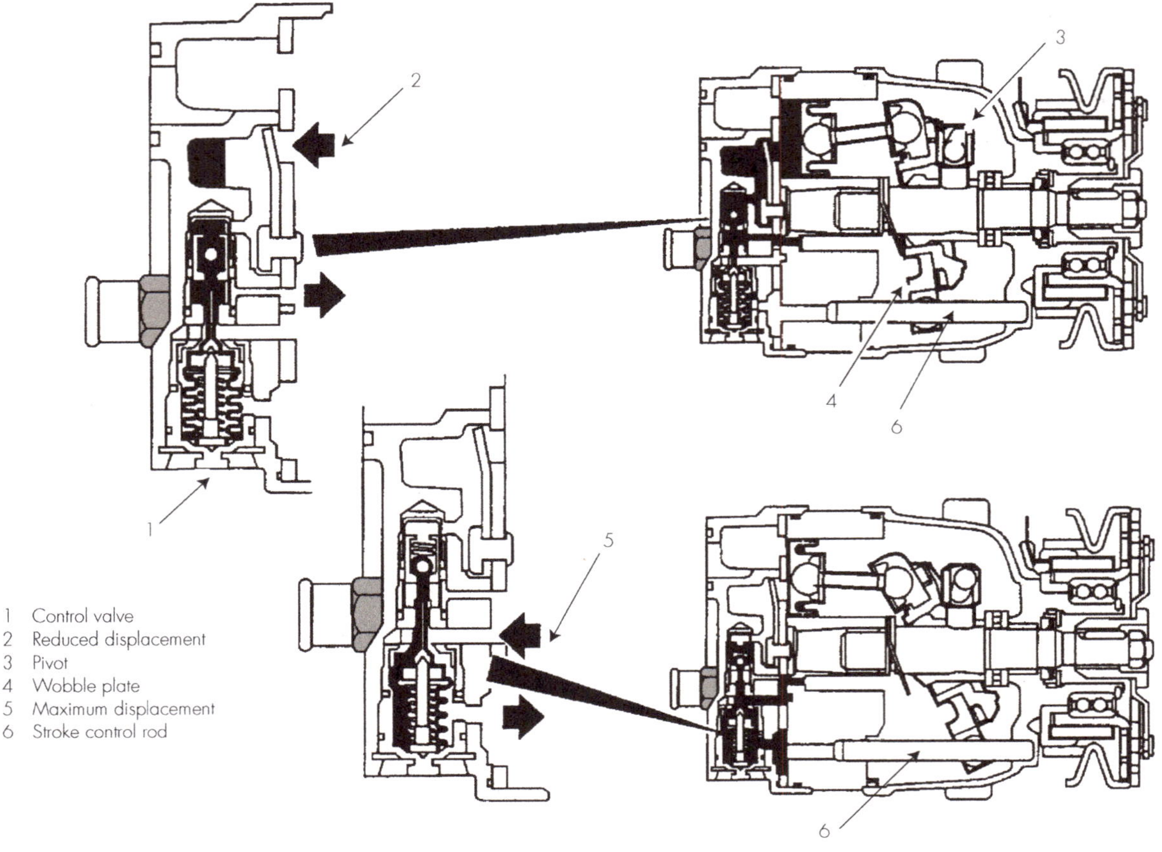

Figure 10.5 Action of the variable displacement compressor
Source: General Motors Holden

The computer module monitors the system operation through various sensors throughout the air-conditioning and engine system. The control module then emits a pulse-width modulated signal that opens and closes the control valve and orifice at different frequencies, thereby controlling pressure under the pistons and therefore the axial plate angle and the stroke of the pistons.

The advantage of the electronic control compressor is that it offers more sophisticated and precise control of the compressor, but it can only be used with a computer-controlled air-conditioning system.

Advantages of variable displacement compressors

The use of displacement compressors lowers the energy required from the engine to drive the system when full power is not required. The displacement can be varied, with the load requirement ranging from 1 to 100 per cent.

This enables the system to operate at maximum efficiency at all times.

Another advantage of a variable displacement compressor is that there is no need to cycle the **magnetic clutch** on and off and in some cases the clutch has been eliminated completely. The variable displacement compressor gives the ability to hold the discharge air from the evaporator at a constant temperature.

Hybrid vehicle compressors

These compressors operate the same internal pumping principles as other compressors except for the following:

- The compressor is like a starter motor: as it is electrically driven. Two hundred volts or more are used to drive the compressor (see Figure 10.6).
- A different oil (ND-type 11 PAG oil) is used to lubricate the internal components. This is due to the electrical laminated windings inside the unit. The oils used in other compressors will delaminate the windings in this compressor.

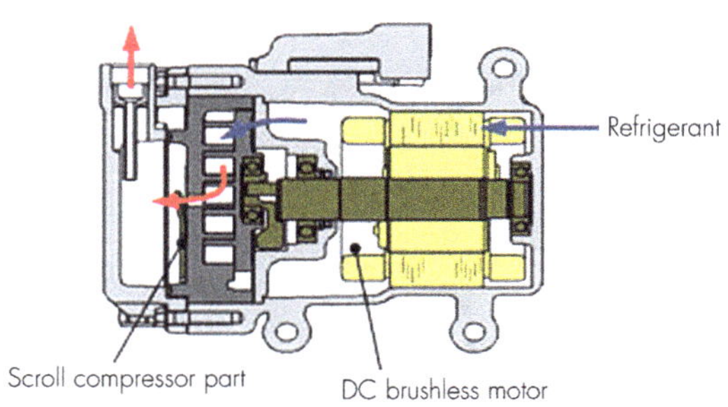

Figure 10.6 Hybrid vehicle electric compressor
Source: Copyright ® 2005 DENSO CORPORATION. All rights reserved

Check manufacturer's specifications for the correct oil for that particular system.

Other system changes include:

- a capillary tube and an ejector placed before the evaporator to lower the refrigerant temperature and compressor load
- a solar ventilation system to keep the interior cooler
- a remote air-conditioning start that can be set for up to three minutes prior to entry so that the vehicle interior is cooled.
- heat that is generated from the exhaust to maintain coolant temperature and interior heating
- the use of an electric water pump.

Compressor drives

The drive belt drives the clutch pulley, which has a **field coil** encased within it. When the field coil is energised, thus creating an electromagnetic field (3 to 4 amps), it pulls the drive plate on to the clutch pulley. The drive plate in turn is keyed to the compressor crankshaft. Therefore, when the field coil is energised, the compressor will start to pump.

The clutch pulley is driven by any of these:

- a single V-belt
- twin V-belts
- a rib belt (serpentine belt).

Compressor drive belt tensioning

There are several ways of tensioning the compressor drive belt. Only the more popular methods will be discussed in this service procedure.

Precautions

NOTE

+ Excessive compressor belt tension will cause a premature failure of the clutch armature pulley bearing. It is essential to align the pulleys for the crankshaft, idler assembly and compressor.

Pulley alignment

Each pulley needs to be aligned with all other pulleys that share the same drive belt. Any misalignment will cause rapid belt wear and damage to the component bearings.

There are three types of belts.

› V-section belts (long used to drive engine accessories)
› V-section cut cog belts
› multi-ribbed belts called 'poly-rib belts' or 'serpentine belts' (Figure 10.7 shows the typical positioning of a serpentine belt).

Methods of adjusting belt tension

The correct belt tension is critical to the lateral or radial loadings placed on the compressor front bearings. There are three methods of belt-tensioning:

1 automatic adjusting of spring pulleys to adjust for belt length or wear
2 adjust the idler pulley by moving it in a direction that tensions the belt (see Figure 10.8)
3 using adjustable components, e.g. alternators.

There are two methods of tensioning for both 2 and 3 above:

› using a tension gauge (see Figure 10.9)
› belt deflection using the amount the belt moves away from its line of natural rest (shown in Figure 10.8, and in Figure 10.10 at A, B and C).

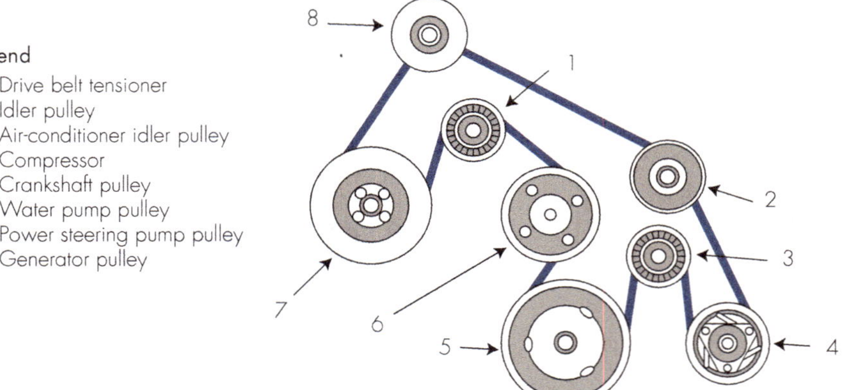

Figure 10.7 Diagram showing serpentine belt positioning
Source: General Motors Holden

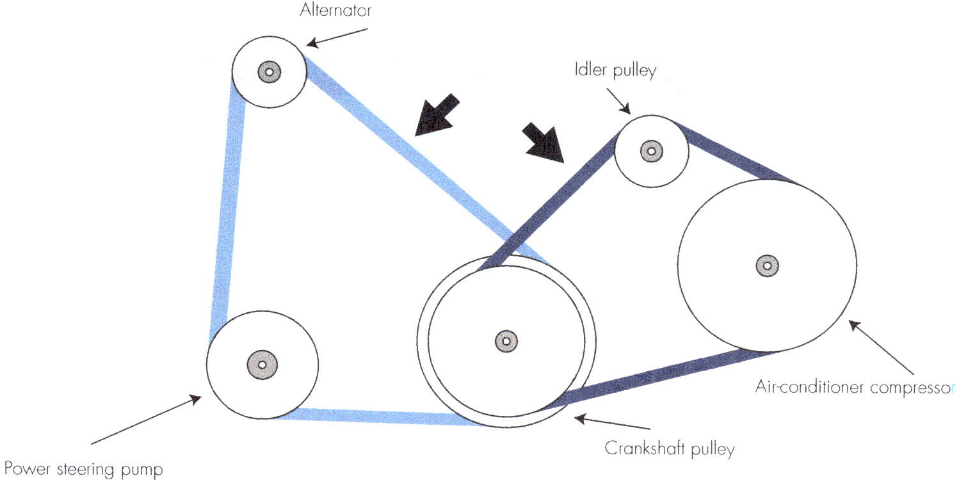

Figure 10.8 Positioning the belt tensioner

Source: General Motors Holden

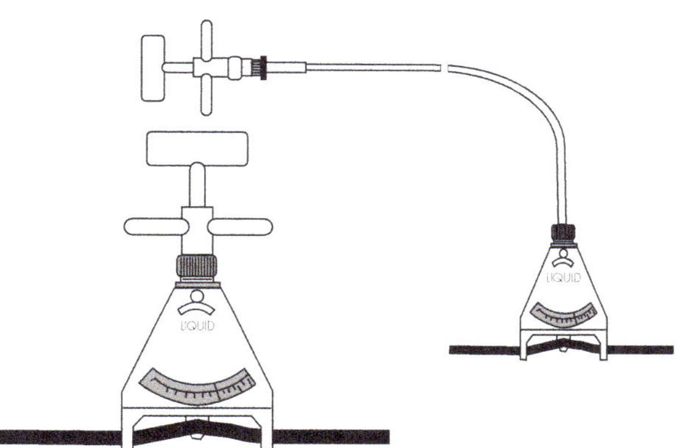

Figure 10.9 Belt tension using a belt tension gauge

Source: General Motors Holden

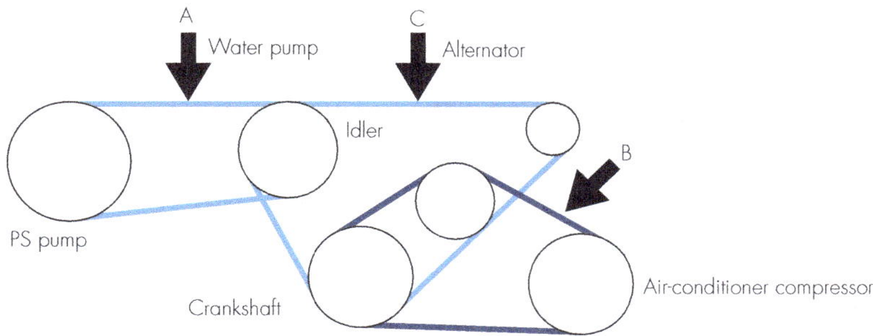

Figure 10.10 Positioning of belt tensioner using the belt deflection method

Source: General Motors Holden

Table 10.1 Air-conditioning belt tensioning chart

Belt type	New belt	Used in service belt
V-belt	490–574N (110–130 lb)	390–440N (90–100 lb)
Serpentine		
3 ribs	490–540N (110–120 lb)	360–390N (80–90 lb)
4 ribs	490–540N (110–120 lb)	360–390N (80–90 lb)
5 ribs	590–670N (140–150 lb)	390–440N (90–I00 lb)
6 ribs or more	670–800N (150–180 lb)	490–540N (110–120 lb)

Tensioning specifications

The pitch and width of belts in the V-belt system are important as they must match both the drive (engine crankshaft) and the driven (engine accessories) pulleys. If the compressor and/or alternator are driven by two belts, they must be replaced as a factory-matched set. This is true even if one of the pair appears to be undamaged.

Over-tightening will result in belt wear and bearing damage, while a loose belt will cause the belt to wear and slip. It is always necessary to re-tension new V-belts after a short run-in period, as they tend to settle into the pulley groove and stretch. Refer to Table 10.1.

When measuring the deflection of the belt, apply a 10 kg force (98N) to the part of the belt that has the longest span between pulleys.

New belt 8–10mm deflection
Used in-service belt 10–12mm deflection

V-belts should never touch or run on the bottom of the pulley groove (see note in Figure 10.11). They are designed to run on the sides of the pulley, and should be replaced if the gap

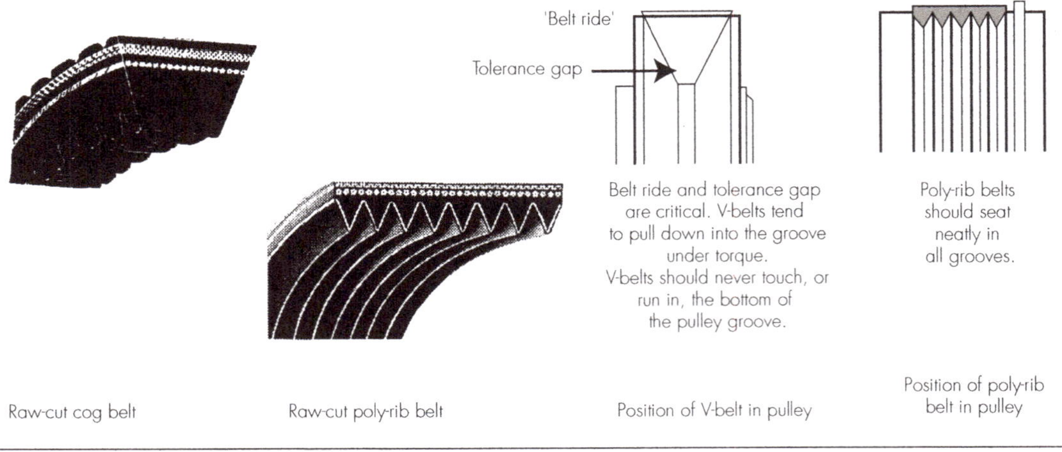

Figure 10.11 Compressor drive belts
Source: Atkins Carlyle Car Parts

between the belt and the bottom of the pulley becomes too small. This gap will disappear under load as the belt will then pull down into the groove.

Many manufacturers are now replacing V-belts with multi-rib belts in order to increase the drive surface of the belt. Most serpentine belts are multi-rib.

The serpentine belt must be replaced with an exact duplicate. The belt's length, width and multi-rib profile are important to ensure correct fit. This system uses a spring-loaded belt tensioner (an 'idler pulley') and so it is not necessary to manually tension this type of belt.

Most manufacturers mark the wear/stretch limit of the belt on the tensioner with an arrow (see 'Pointer' in Figure 10.12). The belt should be replaced when the arrow reaches this point. It is important to note that serpentine belts, unlike V-belts, use the back of the belt as well as the rib side to drive accessories. Therefore, this should be carefully checked for signs of wear as well as the multi-rib side.

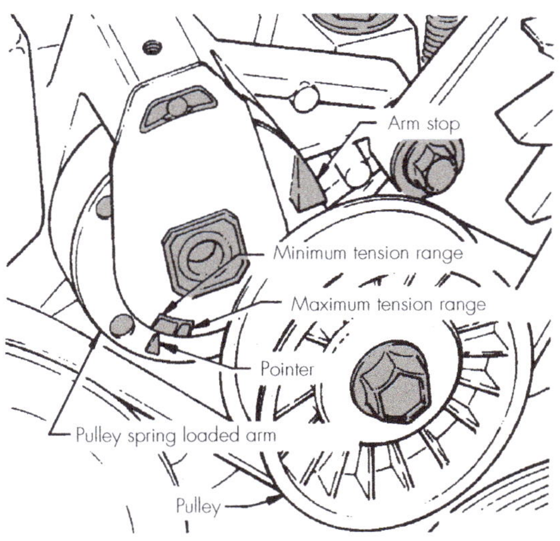

Figure 10.12 Compressor belt tensioning length indicator
Source: General Motors Holden

Compressor repairs and diagnosing problems

Many simple compressor repairs can be carried out during a routine service. These repairs include checking and adding oil, checking the crankshaft seal for leaks and checking the clutch assembly for air gap and coil resistance. More complex repairs are usually not attempted, as the high cost of rebuilding a compressor, compared to the price of replacing it with a new one, means that this is not economically viable.

The 'Service Procedures' section in this text provides information on the various service or repair procedures for a number of common compressors.

Broken or not sealing discharge/suction valves in compressors are not uncommon. The first indication of failure is either a higher than normal low-side (suction) pressure, accompanied by a lower than normal high-side (head) pressure with the gauges fluctuating. Many compressors continue to work adequately with valves that do not seal completely – the only indication of the fault is the flickering gauge. It is only when system performance is affected that repairs need to be carried out.

When you diagnose a problem in a compressor, you need to consider the cost of the repair, the price of the new parts and the possibility of a warranty claim against you if there are any further problems. You then need to weigh all this against the cost of a new or reconditioned compressor with the manufacturer's warranty.

Chapter-revision questions

SELF-CHECK QUESTIONS

1. What are the two main functions of a compressor?
2. What are the three main designs of air-conditioning compressors?
3. How is the drive connected and disconnected to the compressor?
4. What symptoms would a faulty reed valve show?
5. How many valves does a scroll compressor use?
6. In a variable displacement compressor, how is the stroke varied?
7. What advantage does a variable displacement compressor have over conventional compressors?
8. What are the three common types of drive belts used by air-conditioning compressors?
9. What is the approximate belt tension of a 5-rib serpentine belt that has been on the vehicle for 20 000 km?
10. What gauge readings would you expect from a compressor with faulty sealing rings?

CHAPTER 11

Expansion valves

Objectives

On completion of this chapter the student will be able to:

- state the location of the TX valve and the FOT in the circuit
- describe their purpose
- identify the major parts of the TX valve
- describe the operation of the TX valve and the FOT
- recognise the two different types of TX valves (internally and externally equalised)
- state other trade names of components that house the TX valves.

This chapter covers some of the essential knowledge and associated skills involved with:

- AURT222670A Service air-conditioning systems – refrigeration theory.

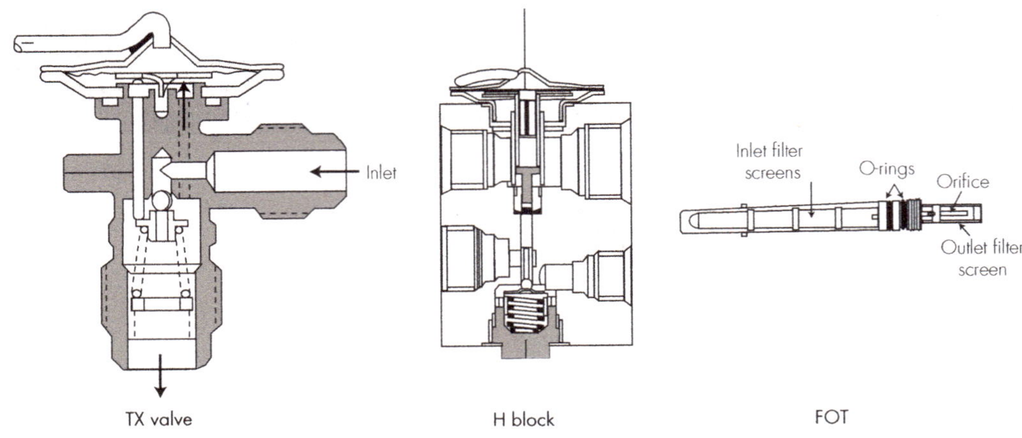

Figure 11.1 Diagrammatic representation of a TX valve, an H block valve and a fixed orifice tube (FOT)
Source: Atkins Carlyle Car Parts

An expansion valve controls the amount of refrigerant that enters the evaporator. This device is known as a thermostatic expansion valve (TX valve). Other expansion valves names are the H block valve and the fixed orifice tube (FOT) (see Figure 11.1). Expansion valves are metering devices and must have 100 per cent liquid to operate effectively.

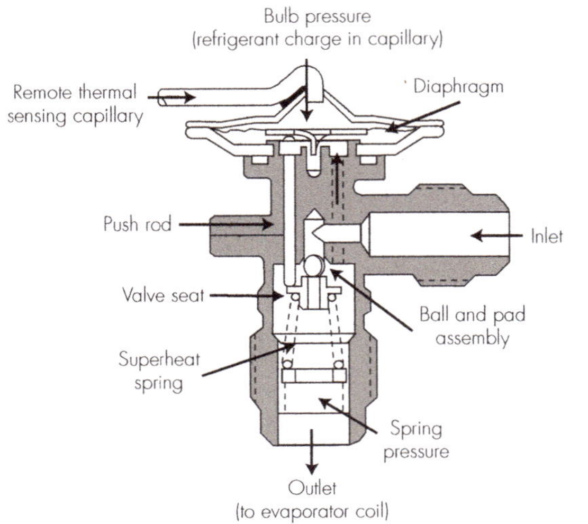

Figure 11.2 Internally equalised TX valve
Source: Atkins Carlyle Car Parts

These valves are all positioned at or near the inlet of the evaporator on the high-pressure liquid line. Their purpose is to control the amount of refrigerant entering the evaporator for the purpose of converting the high-pressure liquid to a low-pressure liquid. This in turn will boil and change state to a vapour. Each type of control valve will be discussed separately in this chapter.

Thermostatic expansion valve (TX valve)

There are two types of TX valve in common use:

- internally equalised (see Figure 11.2)
- externally equalised (see Figure 11.3).

The operation of these valves will be discussed together, but the differences in the equalising actions will be discussed separately.

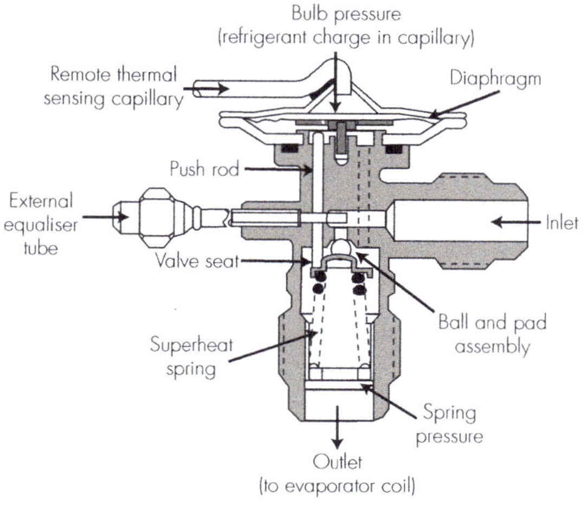

Figure 11.3 Externally equalised TX valve
Source: Atkins Carlyle Car Parts

TX valve parts

Figures 11.2 and 11.3 show the positions of these parts internally and externally equalised TX valves:

- valve seat
- valve diaphragm
- ball and pad assembly
- superheat spring and adjuster
- remote thermal sensing capillary and coil or bulb
- push rod.

Operation of the thermostatic expansion valve

The high-pressure liquid line directs liquid refrigerant (via the inlet) into the TX valve where it meets a ball valve resting on its seat to restrict the flow of refrigerant. The push rods rest on the valve seat and are activated by the diaphragm in order to open or close the ball valve. The ball valve movement is controlled by three forces:

- The pressure difference within the evaporator pushes the diaphragm upwards, which tends to keep the valve closed. A pressure rise within the evaporator closes the valve thus restricting suction pressure.
- The superheat spring tension pushes the valve against the seat, which also tends to keep the valve closed.
- The pressure of the liquid refrigerant in the remote thermal sensing capillary. This senses the temperature as the evaporator outlet expands and pushes the diaphragm down against evaporator pressure and tends to try to open the valve. Temperature rises at the evaporator outlet open the valve. This allows more refrigerant into the evaporator.

The purpose of the thermostatic expansion valve is to balance all three processes and keep enough refrigerant in the evaporator to maintain approximately 0 to 2°C at the evaporator outlet.

Remote thermal sensing capillary tube and coil or bulb

The liquid refrigerant in the capillary and coil is usually the same type that was originally used in the system. The same refrigerant is used in the capillary because it has the same characteristics as the refrigerant in the system for any given pressure/temperature.

Under normal conditions the liquid refrigerant starts to boil as it passes through the evaporator assembly and picks up heat from the passing ambient air. The refrigerant will be 100 per cent vapour by the time it leaves the evaporator and could take on some superheat in the process. The remote thermal sensing capillary coil is clamped onto the suction return line at the evaporator outlet, where it can sense the warmer temperature of the vapour as it leaves the evaporator. The temperature (and therefore the pressure of the refrigerant) will increase inside the capillary. The pressure on the top of the diaphragm will increase, pushing the push rods down to open the ball valve, thus allowing more refrigerant to flow into the evaporator.

If the ball valve is closed, not enough refrigerant will flow into the evaporator. The suction pressure will be low and the return line temperature will be raised (4 to 10°C). This will cause a pressure rise in the capillary, which pushes against the diaphragm to open the ball valve, thus allowing more refrigerant to flow. In turn, this will increase the flow of refrigerant through the evaporator and lower the temperature of the low-pressure return line.

When the ball valve is opened, too much refrigerant will flow into the evaporator, causing the low pressure to be higher and the return line temperature to be lower (−1 to −3°C). This will create a positive pressure under the diaphragm as well as a lowering of pressure above the diaphragm. This will result in the ball valve closing.

The TX valve has two main functions:

› to **throttle**
› to modulate.

Throttling

The refrigerant entering the TX valve is a high-pressure liquid, but it leaves the TX valve as a low-pressure liquid. This drop in pressure is achieved without changing state.

The TX valve is a restrictor that throttles or restricts the flow of the refrigerant.

Modulation

The TX valve is designed to allow only the correct amount of refrigerant to enter the evaporator – at all times and under all conditions – in order to provide the correct degree of cooling. The amount of refrigerant needed will vary according to the heat load situation, and so the TX valve needs to move between the minimum restriction and the maximum restriction positions (modulate) as required.

The word 'modulation' means, in this case, changing the amount of liquid refrigerant that enters the evaporator in response to the load placed on the system. As the heat load increases, more refrigerant flow is required. As the heat load decreases, less refrigerant is required and the valve can start to close down.

This flow-control action of the TX valve maintains proper refrigerant metering under all heat load conditions. When performance testing the system this modulation action can be seen in the suction (low) pressure changing as the system stabilises thus testing the valve operation.

Superheat

Liquid refrigerant vaporises at approximately −26 to −29°C and remains cold even after all the liquid has vaporised.

The vaporised refrigerant continues to flow through the evaporator, absorbing more heat as it goes and thereby becoming superheated. In other words, the temperature of the refrigerant rises above the temperature at which it would normally vaporise.

All TX valves are adjusted to operate under normal superheat levels for the particular system and specified refrigerant type. When the TX valve is being replaced, it is important to ensure that it has the correct superheat range and setting.

Equalisers

It has already been noted that TX valves are either externally or internally equalised. This describes the provisions made for exerting pressure under the diaphragm.

Internally equalised

See Figure 11.2 for a diagram of an internally equalised TX valve.

Note that the push rod passes through drillings below the ball valve. This places pressure on the underside of the diaphragm, which works in conjunction with the pressure exerted on top of the diaphragm by the capillary and coil to modulate the valve.

Externally equalised

See Figure 11.3 for a diagram of an externally equalised TX valve.

This type of TX valve functions in a similar manner to that of the internally equalised valve, except that a fluid pipe is connected from the underside of the diaphragm to the evaporator outlet in the suction line. In this case the suction return line pressure and the temperature actually control the TX valve. The externally equalised TX valve is used on systems that have larger evaporators. This is because the pressure drops across the unit.

H block valve

Figure 11.4 shows an H block valve. The evaporator inlet (high-pressure liquid) and the suction return line (low-pressure vapour) both pass through the same valve assembly. This enables the H or block valve to do away with the capillary tube and coil. This type of TX valve is easier to service, as it is usually mounted on the firewall and not under the dash.

The suction line temperature causes the pressure difference in the sealed sensing bulb

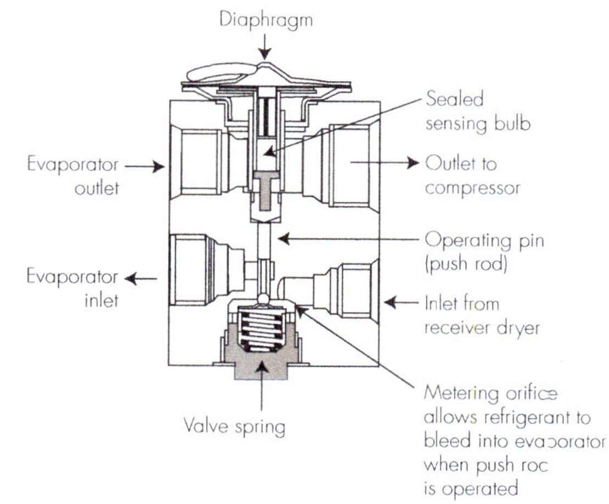

Figure 11.4 The H block TX valve

to open or close the flow of refrigerant into the evaporator through the ball valve.

Fixed orifice tube (FOT)

This device replaces the TX valve. It is located in the high-pressure liquid line usually near the evaporator inlet, but can also be found further back in the line (see Figure 11.5).

There is a fixed drilling (orifice) inside the FOT through which the refrigerant must pass before it can enter the evaporator. Unlike the TX valve, the FOT has no moving parts and is not adjustable. It does not modulate. However, there are various size drillings to suit different system requirements. See Chapter 6 (retrofitting) instructions on FOT sizing.

If the heat load on the system alters and extra cooling is required, the compressor cycle times or piston effective stroke (varied stroke) must be altered in accordance with the increased need.

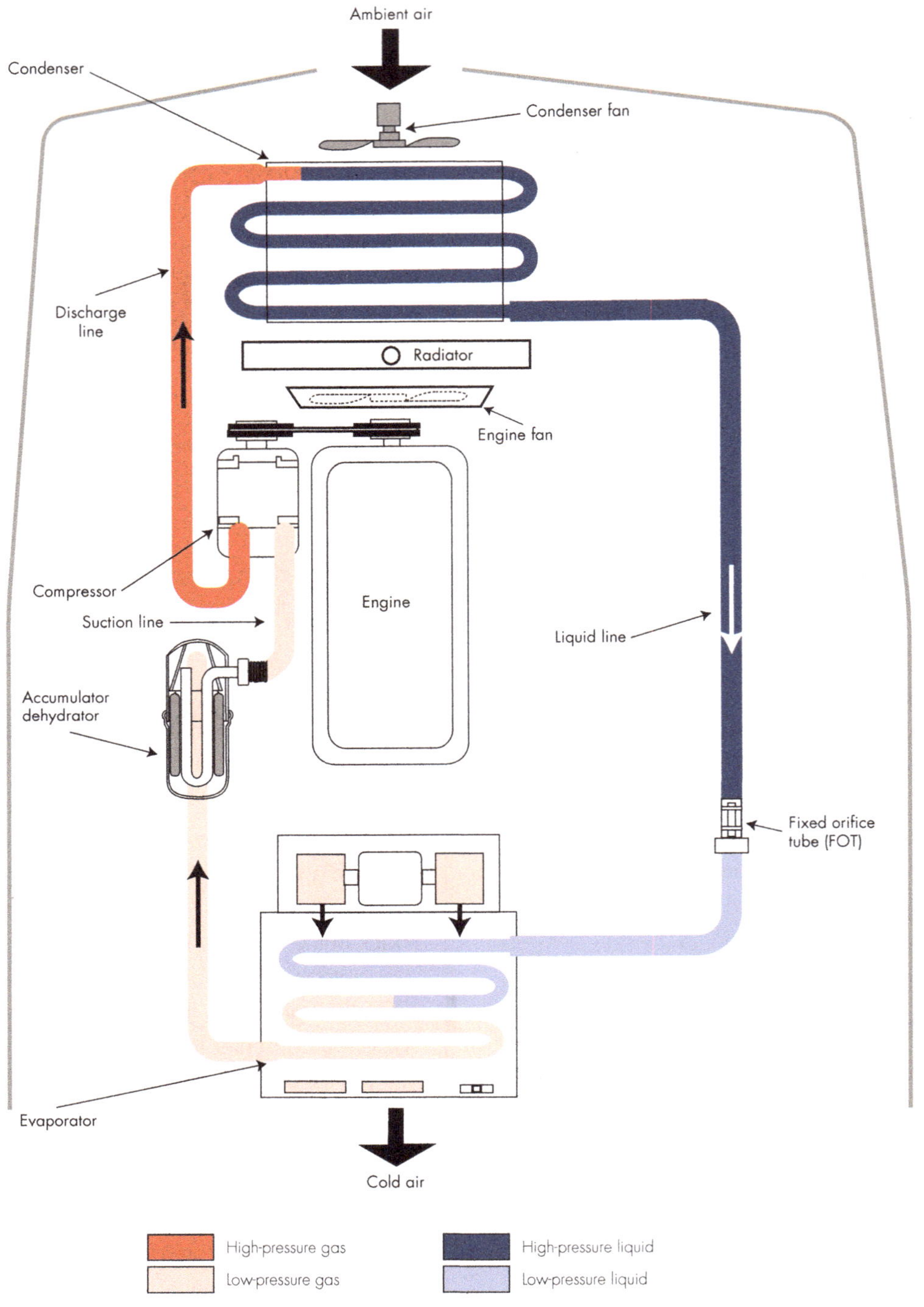

Figure 11.5 Fixed orifice tube (FOT) accumulator system
Source: Atkins Carlyle Car Parts

The FOT system does not have a receiver dryer on the high-pressure side, but it does have an accumulator on the low-pressure return line. After-market FOT's are available with dual flow capabilities. These valves are able to alter the refrigerant flow at lower compressor rotational speeds to maintain cooling and then increase the flow at higher rotational speeds. This reduces flooding the evaporator at idle.

NOTE

+ The air-conditioning system should be operated each week for at least five minutes (even during the winter months when it is not normal to use an air conditioner). This is to prevent vital parts of the TX valve and FOT from corroding or sticking, and to lubricate all moving parts of the system.

Chapter-revision questions

SELF-CHECK QUESTIONS

1. What is another name for a thermostatic expansion valve?
2. What are the two most common thermostatic expansion valves?
3. What is the state of the refrigerant as it enters the TX valve?
4. What two components can form a seal to control pressure within the TX valve?
5. What temperature is the refrigerant as it leaves the evaporator?
6. What is the state of the refrigerant as it leaves the evaporator?
7. What are the two main functions of the TX valve?
8. What type of device can be used instead of the TX valve?
9. How does a FOT system cope with the change in heat load in the evaporator?
10. Why should the air-conditioning system be operated every week even in winter?

CHAPTER 12

Electrical and other system controls

Objectives

On completion of this chapter the student will be able to:

- follow a simple electrical diagram of an air-conditioning circuit
- state the purpose and types of fuses
- name the basic components in the electrical circuit and describe their purpose
- state the purpose of vacuum controls and how vacuum is controlled.

This chapter covers some of the essential knowledge and associated skills involved with:

- AURE218708A Carry out repairs to single electrical circuits
- AURT222670A Servicing air-conditioning systems.

Electrical system operation

The basic air-conditioning electrical circuit is a series circuit. In its simplest form it uses a fuse, a fan speed-control switch, a **thermostat** and a compressor clutch coil. The blower fan is connected by itself away from the main circuit (see Figures 12.1 and 12.2). If any one part of a series circuit is faulty, the whole circuit will not work.

Modern vehicles, with engine management and other sophisticated controls that maximise fuel economy and driveability, use an air-conditioning electrical circuit that is a little more complicated than those shown in these two diagrams. However, a good understanding of the basic electrical circuit is essential if the air-conditioning technician is to accurately diagnose problems.

Figure 12.3 is typical of a modern vehicle's air-conditioning circuit.

Note that the circuit contains a:

- cooling fan control circuit
- compressor control circuit. This circuit is protected against the risk of pressures becoming too high or low because of a lack of refrigerant.

Figure 12.4 shows an air-conditioning system controlled by electronic control unit (ECU).

Each of the components in the circuit is discussed and their operation in the circuit explained in the following sections. They are numbered (1–15) for easy reference to Figure 12.3. Please refer to the relevant numbered component to understand how the circuit operates.

1 Fuses

Every circuit in modern motor vehicles must have some form of circuit protection, in case of a fault.

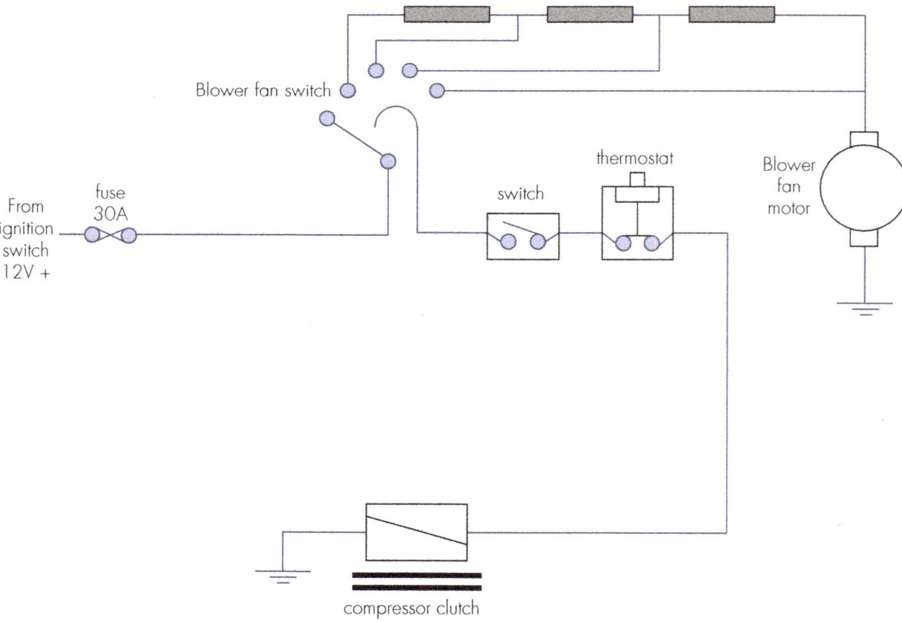

Figure 12.1 A simple air-conditioning circuit before 1985

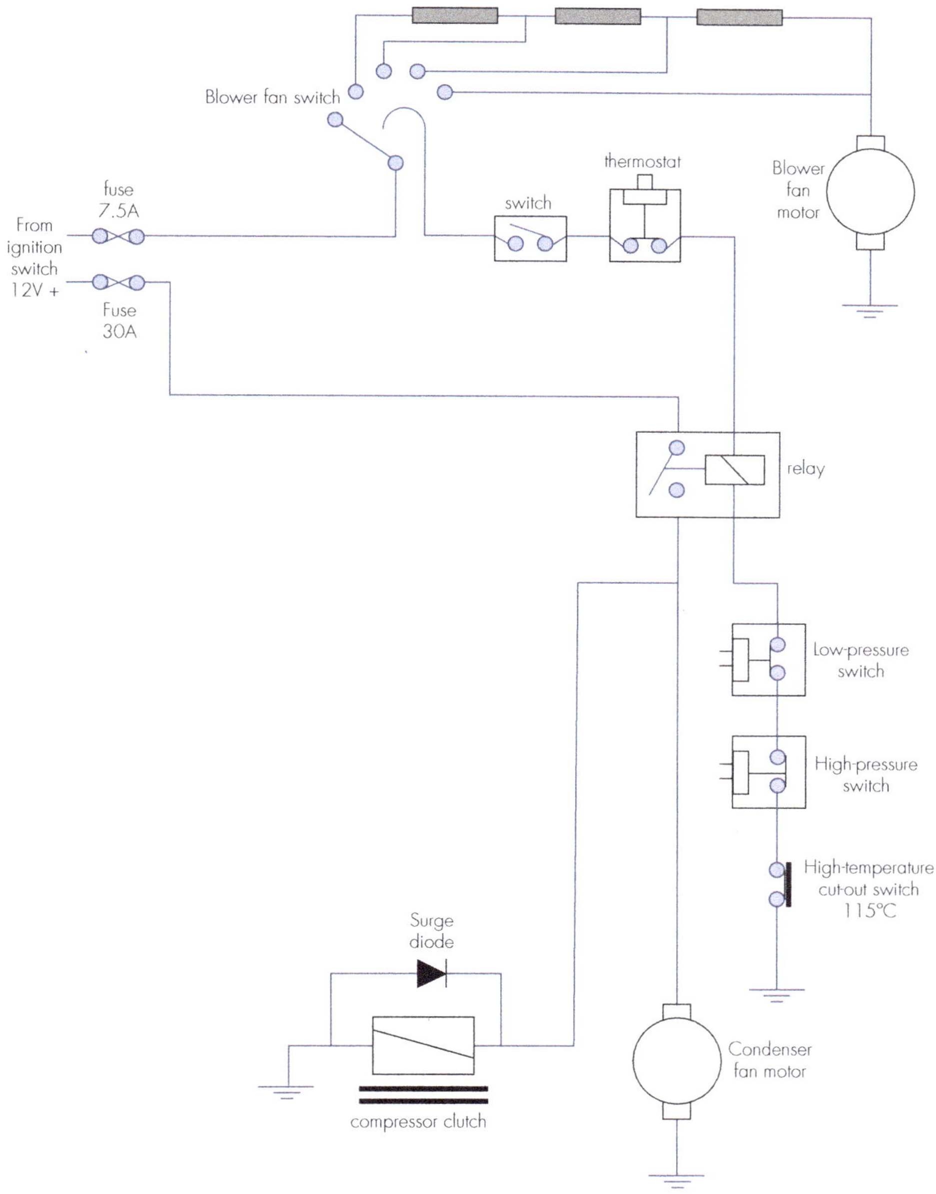

Figure 12.2 A simple air-conditioning circuit as used from 1986 onwards

The air-conditioning circuit may use its own fuse, or it may share a fuse with another circuit.

Fuses are rated in amperes or amps. If one should fail, only a fuse of the same amp rating should be used to replace it.

Fuses are designed to blow or burn out if the current flow in the circuit exceeds the rating on the fuse. This prevents the circuit wiring from burning and any sensitive components from being damaged. If a blown

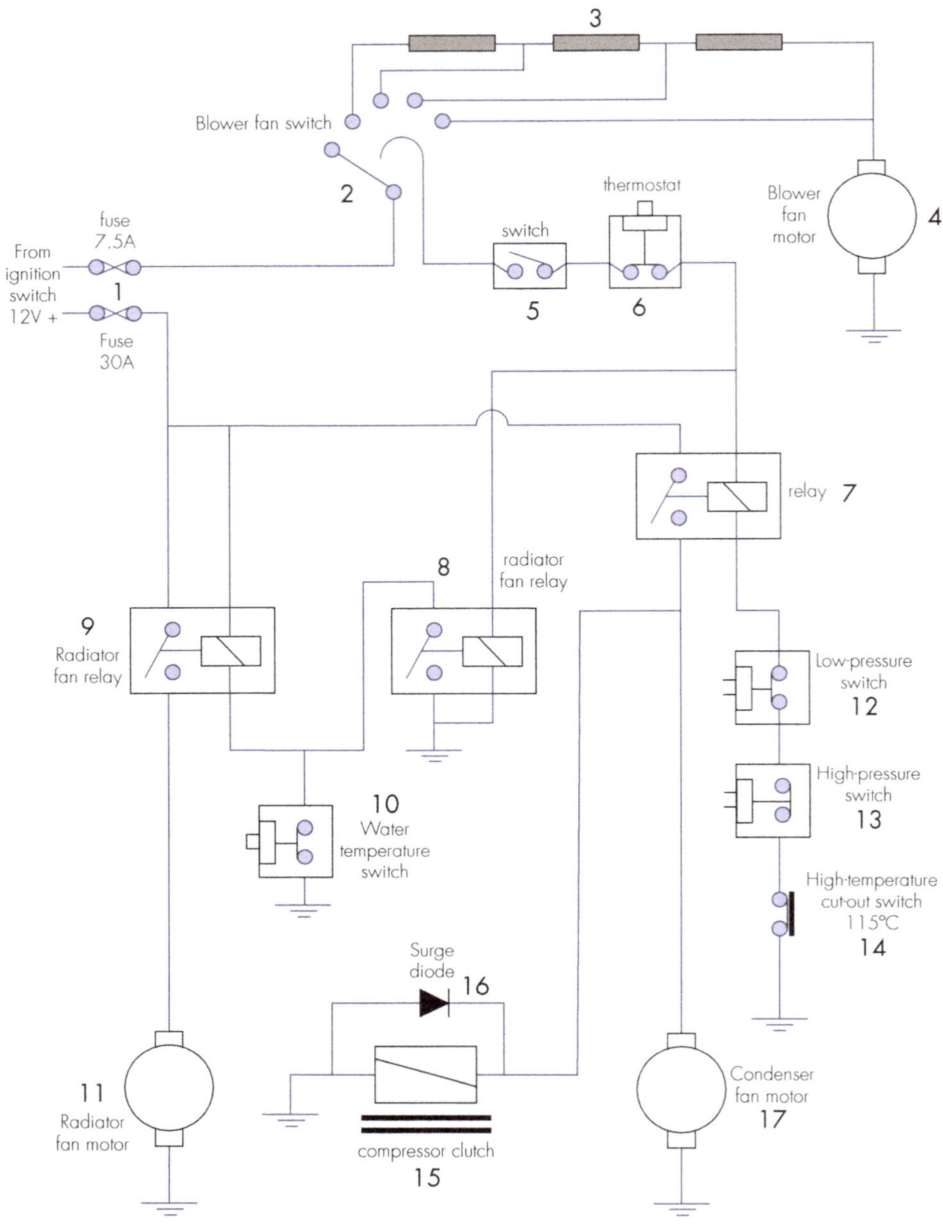

Figure 12.3 A typical air-conditioning circuit as used on current vehicles

air-conditioning fuse is discovered, the wiring and electrical components must be checked to find the cause of the excessive current flow. This may be caused by a wire rubbing through its insulation, or by a short circuit in a fan or blower motor if they are on the same fuse circuit. Fuses can also fail due to thermal (heat) fatigue or age and a new fuse will rectify the

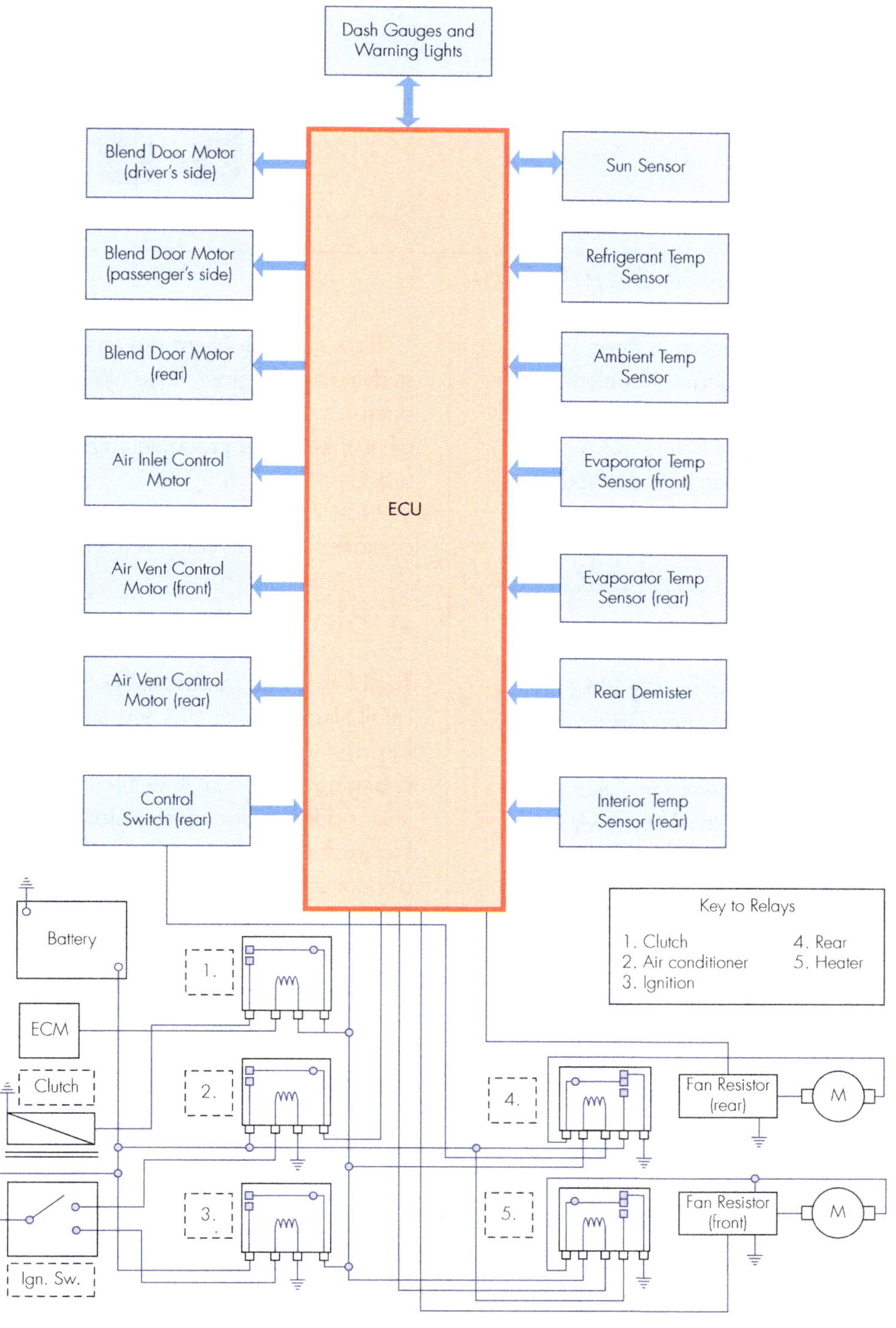

Figure 12.4 Air-conditioning system controlled by electronic control unit (ECU)

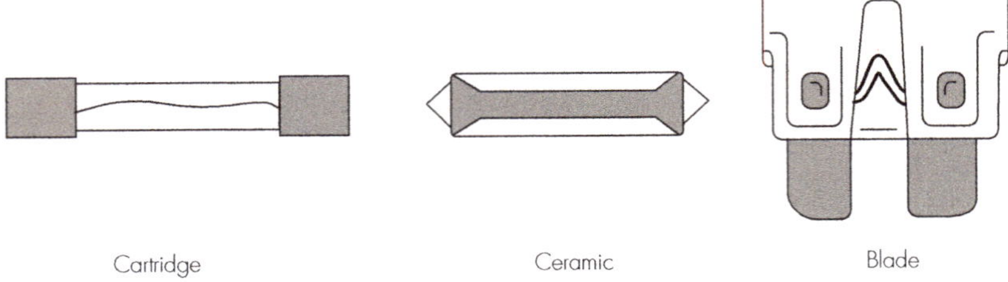

Figure 12.5 The three different types of electrical fuses

problem. Fuses are available in three styles, which are not interchangeable – only the fuse designed for the vehicle should be used. Study the three types of fuses in Figure 12.5 and note that the most common fuse on modern vehicles is the blade type.

2, 3 Resistors and blower fan control

The blower fan switch and resistor block in the circuit diagram control the speed of the blower fan by switching across resistors and varying the current flow to the blower fan. Study the circuit diagram and note that if the switch is in the first position, the current flow is available to the air-conditioning switch and to the first resistor. The current is reduced after it has moved through the first resistor and is further reduced after moving through the second and third resistors, so the fan will rotate at its lowest speed.

If the switch is moved to the second position, the current only has to move through the second and third resistors and so second speed is achieved. In the third switch position, current only has to flow through one resistor. The fourth position or high fan speed is achieved by the switch bypassing all resistors and allowing full current flow to the fan motor.

Blower fan problems are easy to diagnose by studying the diagram. If the blower fan does not operate on low speed, but operates normally on other speeds, the probable cause is a faulty first resistor.

Note that the air-conditioning system will not operate if the blower fan is not switched on.

4 Blower fan and motor

The blower and fan assembly consists of a small electric motor that drives a squirrel cage fan. The various speeds are achieved by changing the current flow through the motor. This is done by the resistor block and the blower fan control switch. The blower fans are not usually repairable, and need to be replaced if faulty. As the motor is usually bolted onto non-conductive material, always make sure that the earth wire is correctly connected. Noisy fans are sometimes a result of leaves trapped in the housing – clearing of these requires the removal of the blower fan.

5 Air-conditioning switch

The air-conditioning control switch only has power to it after the blower fan control is switched on. The switch is usually located on the dash panel where, when selected, a light illuminates to indicate that it is on.

6 Thermostat

The thermostat is an electrical switch controlled by the temperature of the evaporator. If the compressor is allowed to remain cycled on at all times, the moisture in the air and the temperature of the refrigerant flowing through the evaporator will soon cause it to freeze over. To prevent this, the thermostat senses the temperature of the evaporator. When the temperature approaches 0°C the thermostat opens its switch, cycling off the compressor by stopping current supply to the electromagnetic clutch at the compressor.

As the temperature rises, any ice that has formed on the evaporator core will melt and drip into the drain tray before leaving the vehicle via the drain tube. The thermostat senses the rise in evaporator temperature and closes its switch, causing the compressor to once again cycle in. A thermostat is sometimes known as a 'de-icing switch'.

Two types of thermostats are commonly used on modern vehicles – the 'bellows' or 'mechanical' type, and the '**thermistor**' or 'electronic' type. The thermistor is replacing the bellows type as it is considerably more sensitive to temperature changes.

Bellows thermostat

The bellows-type thermostat uses a capillary tube inserted into the evaporator. The capillary tube is usually probed into the evaporator fins to at least 25 mm. This is necessary in order to record an accurate indication of refrigerant temperature.
The capillary tube is filled with the same refrigerant that was originally used in the system. This is a heat-sensitive refrigerant that expands when heated, pushing against a spring-tensioned diaphragm. This closes the thermostat contacts, and the compressor is switched on. The evaporator temperature begins to drop and the refrigerant cools and contracts. Eventually the contacts open, cycling the compressor out again. Most thermostats are adjustable (an operation that is covered in the service section of this book) and some are operated mechanically by the driver of the vehicle, thus allowing controlled temperature choices. The spring tension can be varied by turning the adjusting cam screw, so that the compressor cycling increases or decreases. This type of thermostat is still used, but is less common than thermistor thermostats.

Thermistor thermostats

The thermistor is actually a resistor whose resistance value is determined by its temperature. Although they may vary in physical appearance, all thermistors have the same general operating characteristics. They are extremely sensitive to slight changes in temperature, and the change in each thermistor's resistance value is inversely proportional to its temperature change. When the temperature decreases, the thermistor's resistance increases; when the temperature increases, thermistor resistance decreases. As the resistance changes, the current flow also changes.

As an example, assume that a thermistor has been installed in the evaporator duct. With air at a temperature of 16°C passing through the duct, the resistance value of the thermistor is approximately 1900 ohms. If the temperature in the duct is 32°C, the resistance of the thermistor decreases to about 800 ohms. If, however, the temperature is decreased to 4°C the thermistor resistance is increased to approximately 2500 ohms. These resistance values are sent to a control unit amplifier that, at a predetermined voltage, interrupts the current flow to the compressor, thus cycling the compressor.

7–11 Relays

There are usually several relays in modern vehicles to control areas such as the air-conditioning clutch, engine cooling and condenser fan operation.

A relay could be described as an electro-magnetic switch. Its main function is to use a small current flow to turn on a larger one, thus preventing the large current flow from being passed across switches and avoiding the risk of arcing and voltage drop throughout the circuit.

Whenever a current is passed through a coil of wire that has been wound around a soft iron core, an electromagnet is produced. This electromagnet is used to attract the spring-loaded arm of a switch, either turning the switch off or, more commonly, on.

Three relays are shown in Figure 12.3. They are numbered 7, 8 and 9.

The *air-conditioning relay (7)* uses the power from the air-conditioning switch to turn on the relay and switch both the compressor clutch and the condenser fan motor on.

The *air-conditioning radiator fan relay (8)* is used to switch on the radiator fan relay, which in turn starts the *radiator fan motor (11)*. This provides extra condenser cooling and ensures that the engine does not overheat.

The *radiator fan relay (9)* is activated by the *air-conditioning radiator fan relay* when the air conditioning is operating, but when it is switched off the radiator fan relay receives its earth from the *water temperature switch (10)*, which is a temperature-sensitive switch that switches on when the temperature rises above a predetermined setting (usually 92 to 95°C).

12, 13 Pressure-sensing switches

Pressure-sensing switches are designed to open when a predetermined pressure has been reached. By law they are required to be fitted to all air-conditioning systems. The pressure switches prevent air-conditioning systems from operating by cutting power to the compressor if the pressures are too low because refrigerant has been lost as a result of leaks, and if the system pressures are too high because of poor cooling at the condenser. There are four designs of pressure switches in use. The pressure ratings for each switch are usually printed on the side of the switch.

The *low-pressure switch (12)* is fitted on the low side of the air-conditioning system, and is set to open when the system pressure falls below 30 to 35 kPa.

The *high-pressure switch (13)* is on the high side of the system and is set to open and stop the compressor from operating if system pressures rise above a set level (2400–3200 kPa, according to the system design).

The *high-side low-pressure switch* is also situated on the high side. It cuts out compressor operation when the system pressure falls below 250 kPa.

The *high/low-pressure switch* is a **binary switch**, which senses both high and low pressures. It is the most common type of switch, and must be fitted to the high-pressure side of the air-conditioning system.

A binary switch may also be a mid/high-pressure switch. The mid-pressure part of the switch turns on the condenser fan in an attempt to limit pressure increase. If this fails and the pressure continues to rise, the compressor is stopped.

Pressure transducers

Pressure transducers (Figure 12.6) are replacing pressure switches in electronically-controlled systems.

They are sealed reference capacitive pressure sensors with two-piece ceramic diaphragms. Their main advantage over the normal type of pressure switch is that the transducer is constantly monitoring the system pressures and sending signals to the control module. The pressure **transducer** is usually combined with the service valve coupling connection.

14 High-temperature (thermal) cut-out switch

This switch is sometimes referred to as a 'superheat switch'. It senses the temperature of the compressor casing and opens the compressor circuit if the rated temperature (115°C) is exceeded. This saves the compressor from damage due to overheating.

15 Air-conditioning compressor clutch

The air-conditioning compressor clutch is used to connect and disconnect the drive between the engine crankshaft and the compressor.

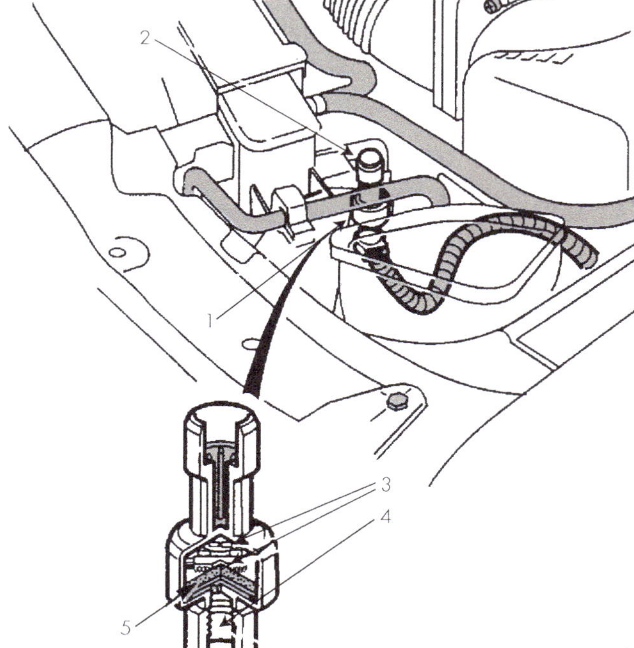

Figure 12.6 A typical pressure transducer
Source: General Motors Holden

The clutch assembly consists of a field coil, a **drive pulley** and a clutch hub or 'shoe'.

The field coil consists of many turns of wire around an iron core, which becomes a strong electromagnet when a current of between 3 and 4 amps is flowing through it. The magnetic field attracts the clutch hub towards the field coil, thus connecting the drive pulley so that it turns the compressor crankshaft. When the current flow stops flowing in the field coil, the magnetic field collapses and strap springs on the clutch hub pull it away from the drive pulley. Shims behind the clutch hub maintain an air gap, usually between 0.4 and 0.8 mm, between the hub and pulley.

When the magnetic field collapses in the field coil a current is generated. This current may damage sensitive electronic components such as the ECU (electronic control unit). In order to prevent this from happening, a surge diode is connected in reverse bias across the field coil. Note its location (16) on the system circuit diagram.

Vacuum controls

Motor vehicle manufacturers often use vacuum mechanisms to open and close vents, demister flaps and heater control taps.

As a vacuum is below atmospheric pressure, this difference in pressure is used to operate vacuum motors.

Vacuum actuators

A vacuum actuator (**servo**) is simply a diaphragm with a return spring behind it and a push rod to apply a force. Some manufacturers now use two-stage or double-diaphragm vacuum units. One side of the diaphragm has a vacuum applied to it from a vacuum source (usually a vacuum control head), while the other side is open to atmospheric pressure (the air pressure around us). The diaphragm will always move towards the area of lower pressure after it has overcome the spring tension. This movement will activate the push rod or 'linkage' so that it opens or closes a flap or door. When the vacuum is redirected or vented by the control head, the spring will move the vacuum motor diaphragm back to the rest position. If the flap door is required to have three operating positions (closed, open halfway and fully open), it needs complicated linkage and two actuators. Manufacturers then use a dual-stage or 'double' actuator (see Figure 12.7).

Vacuum source

The engine's inlet manifold is a ready source of vacuum. However, vacuum can vary and there may even be times (as when the vehicle is under full acceleration) when no vacuum is available. A reserve vacuum tank and a one-way check valve are used to prevent doors and flaps from opening and closing when the vehicle accelerates. The reserve tank stores a ready supply of vacuum that is replenished every time the throttle is closed or almost closed.

Check valve

The check valve prevents the vacuum from being lost back into the engine when the vehicle is under acceleration.

A check valve can be tested by connecting it to a vacuum pump. It should operate in only one direction. Check valves should always be fitted so that the flow is away from the vacuum source. Check valves are not repairable and must be replaced if found to be faulty.

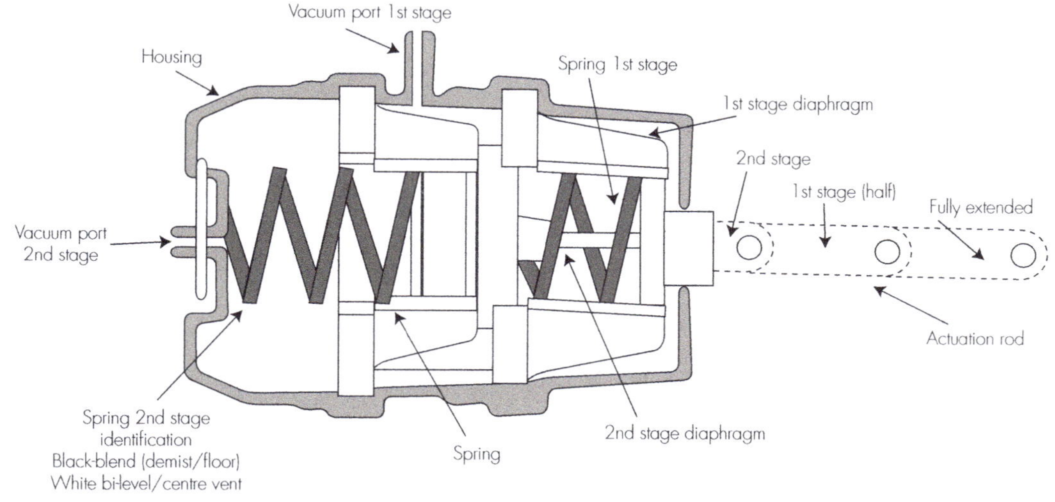

Figure 12.7 A dual-stage (or double-vacuum) actuator
Source: General Motors Holden

Restrictors

Some vacuum systems use restrictors in vacuum lines to delay or slow the operation of vacuum motors. Restrictors are simply connectors with small holes that control vacuum flow. Because the holes are small, they are prone to blocking and they should therefore be checked to ensure that vacuum can flow in both directions.

Other system controls

There are many controls that improve driveability, protect components from damage or help to ensure environmental integrity. Not every vehicle uses them all, but they are becoming increasingly common. Some control compressor operation, while others control refrigerant flow.

Compressor and driveability controls

Idle cut-out

This switch is used on smaller capacity engines. Because an air-conditioning compressor places a heavy demand on these engines, this switch ensures that compressor operation is stopped whenever the engine is idling. This prevents rough idle and possible cutting out of the engine.

Overheat cut-out switch

This switch interrupts compressor operation if the engine is sensed to be overheating. A working compressor places an extra heat load on an already over-heating engine, especially considering that heat is already being radiated from the condenser.

Ambient temperature switch

When the ambient temperature is very low, e.g. below 3°C, this switch prevents operation

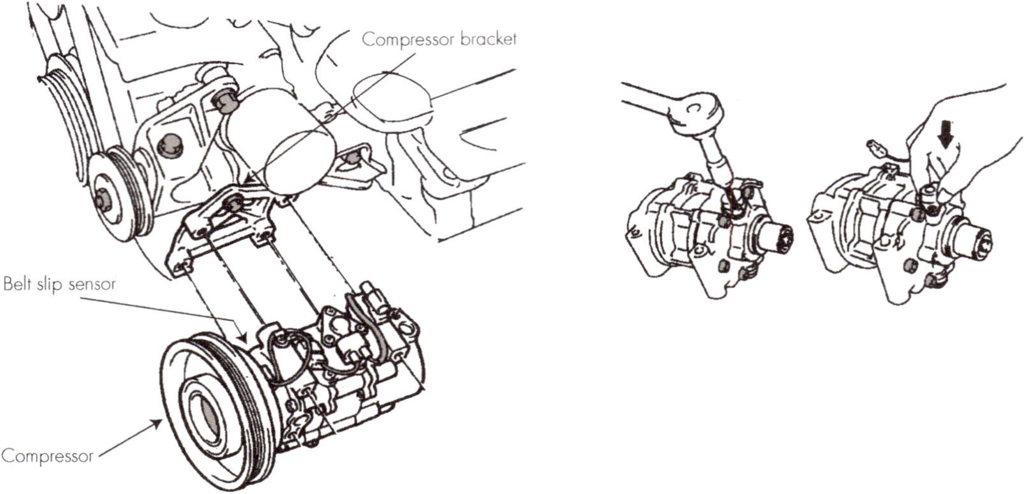

Figure 12.8 The belt slip sensor: position and fitting

of the compressor in non-climate control systems. It is usually located in front of the radiator, in the evaporator air inlet duct or in any other location where it can sense the outside air temperature.

Wide-open throttle (WOT) cut-out switch

The wide-open throttle cut-out switch is common on many smaller cars with low-powered engines. It is usually located on the accelerator pedal and activates a relay which interrupts the compressor clutch circuit when the throttle is wide open. This improves engine performance and makes for quicker acceleration.

Belt slip sensor

A **belt slip sensor** is sensor in the air-conditioning compressor measures the compressor speed and compares it to the crankshaft and engine speed. Any variation in the calculated speed will be because of belt slip. This will cause the electronic control unit to open circuit the compressor. This means that a loose drive belt may cause inoperation of the air-conditioning system. Figure 12.8 shows the location of this sensor.

Reverse operation switch

This switch engages the air-conditioning compressor whenever reverse gear is selected, even if the air conditioning is not switched on. This allows the refrigerant and the oil to circulate and keep all seals and O-rings lubricated, especially during the winter months when it is likely that the air conditioning will not be used. As a side benefit, the cycling of the compressor can also slow the idle speed to smoothly engage the reverse gear.

Idle-up actuators

These can be electrically- or vacuum-actuated units on older vehicles that are fitted with a carburettor. They raise the engine idle speed to compensate for the extra engine load when the compressor cycles in. They can be adjusted to set the idle speed according to the load.

On vehicles with fuel injection and engine management, the electronic control unit will actuate an idle stepper motor or idle air control (IAC) valve to redirect more air around the throttle valve, thus increasing the idle speed according to engine load. These types are not usually adjustable.

IMPORTANT NOTE

+ Always check the manufacturer's workshop manual to determine which sensors and control devices have been fitted to the vehicle. This will make diagnosis easier.

Chapter-revision questions

RESEARCH

Using a late-model vehicle workshop manual, find the section for wiring diagrams and locate the air-conditioning section. On the diagram, trace the wiring from the ECU to the compressor clutch. This may go through a relay first, so make note of the wiring colour, then check the vehicle wiring to confirm your findings.

EXERCISE

1. Test the current draw on a compressor clutch coil as outlined in the service procedures for that particular compressor.
   a. The current draw is ______________
   b. Is this within specifications? ______________________ **(Yes/No)**
2. Test the thermostatic switch as described in Service Procedure 7 on page 208.
   a. Is it within specifications? ______________________ **(Yes/No)**
   b. Adjust if the switch allows for adjustment.

SELF-CHECK QUESTIONS

1. What are the three types of fuses found on modern vehicles?
2. What happens if there is a break in a series circuit?
3. How are fuses rated?
4. What is the function of the air-conditioning thermostat?
5. At what pressure does the low-pressure switch open?
6. At what pressure does the high-pressure switch open?
7. What is a superheat switch?
8. What allows vacuum controls to still operate even though the engine is switched off?
9. On some vehicles the air-conditioning compressor is switched on when reverse gear is selected. Why?
10. What is the current draw of a compressor field coil?

CHAPTER 13

Engine cooling, heating and air-conditioning ducting

Objectives

On completion of this chapter the student will be able to:

- state the purpose of the radiator pressure cap
- test a radiator pressure cap
- identify the purpose of the engine thermostat
- describe the different types of cooling fan
- understand the need to use radiator additives
- name the various sections of the core ducting assembly
- identify the purpose of each section.

This chapter covers some of the essential knowledge and associated skills involved with:

- AURT202170A Inspect and service cooling systems.

Engine cooling system

It is important for the air-conditioning technician to have a good understanding of vehicle cooling systems, as problems with the cooling system may affect the operation of the air-conditioning system or even cause it to cease operation.

This chapter discusses the main components of a cooling system. Their correct operation must be checked before attempting to diagnose air-conditioning problems.

There are two methods of cooling engines: by air and by liquid. The liquid cooling system is by far the most common, and is the system that will be discussed in this chapter.

Radiators

The radiator can be likened to a condenser in that it is the medium for the heat exchange between the coolant and the ambient air.

There are two types of radiator in common use: vertical flow and cross flow. Vertical flow is now only used on larger vehicles such as light and heavy trucks, while cross flow is the most common design in light vehicles because of its low height, which enables vehicle designers to produce vehicles with low frontal styles.

Both radiator designs incorporate two radiator tanks, a series of cooling tubes and rows of cooling fins between the tubes. The hot coolant enters the radiator tank and flows through the tubes to the other tank. The heat is transferred from the coolant to the tubes and then to the cooling fins. These increase the surface area of the radiator. Heat is then transferred to the passing ambient air by the effect of the vehicle's movement or by fans that draw air through the fins.

Checks to be carried out:

- For leaks – done using a radiator pressure tester and pressurising to the same pressure that is stated on the cap. If the pressure drops and no leaks are evident in the radiator, inspect around the engine and coolant hoses for leaks. There is a reason for the drop in pressure.
- The coolant level – done when the system is cold. The cap is removed or the coolant level is inspected through the clear overflow container.
- Annual cleaning of the condenser and radiator to maintain correct airflow.

Radiator pressure caps

A radiator cap is used to contain the coolant in the radiator, but it also has an important secondary function, which is to pressurise the cooling system. By placing the cooling system under pressure, the boiling point of the coolant is increased by some 2°C for every 10 kPa of pressure. This means that a cooling system with a pressure cap rated at 110 kPa will boil at 122°C instead of 100°C. Pressure caps usually have a pressure of 90–150 kPa. A radiator cap is not to be removed from a hot radiator system. Allow at least 15 minutes for the engine to cool before attempting to remove the radiator cap slowly.

The radiator cap also uses a vacuum valve to allow air or coolant back into the radiator from the overflow container, as the system cools and the water molecules contract. This prevents cooling hoses from collapsing or sucking in.

Most modern systems use an overflow tank or header tank to allow for the expansion and contraction of the coolant.

An overflow tank is not pressurised, and the radiator cap is fitted to the radiator. When a header tank is used, the radiator cap is fitted to the header tank, which means that this tank is pressurised and the radiator can only be filled at the header tank. The radiator cap should be regularly inspected and pressure tested using a radiator cap pressure tester as the air-conditioning system places extra loads on the cooling system, especially over the summer months, and a faulty pressure cap will lead to cooling system failure. Remember that a system not holding the recommended pressure can cause cavitation and early pump impeller failure.

Thermostats (the cooling system's TX valve)

A thermostat is a coolant flow control valve designed to prevent the coolant from moving to the radiator when the engine is cold. The thermostat is usually fitted in front of the radiator inlet, and located in a housing in the cylinder head. Some vehicle manufacturers also fit them to the return hose at the water pump inlet to reduce thermal shock in the radiator when the thermostat opens.

Thermostats use wax pellets that expand and open the valve when it is heated. A spring closes the valve when the system cools down. The coolant in the engine is still able to circulate through a bypass hose or passage in the cylinder head. A 'jiggle pin' in the thermostat allows trapped air to bleed out.

Faulty thermostats can cause engines to overheat when they are stuck closed causing overheating, and become slow to warm up when they are stuck open. Auto climate control units will not operate when the engine coolant temperature is cold. This will delay the system operating and will also cause extra fuel usage and an increase in pollution levels. Thermostats usually have their opening temperature stamped on them, and the replacement thermostat must be set at the specified opening temperature. A quick method of testing thermostat operation is checking how long it takes for the gauge to start to move and then to reach operating temperature. It should start to move within a short time (30 to 60 seconds) and reach operating temperature within 2 to 4 minutes.

Water pumps

The cooling system water pump is not a pressure pump but a displacement pump, i.e. it moves or circulates the coolant around the engine cooling passages.

The water pump is driven by one of three methods:

- engine drive belt
- timing belt
- electric motor.

On the other end of the drive mechanism is a rotating impeller, which throws the coolant outwards with a centrifugal action, forcing it to move through the engine.

The most common problem found in water pumps is a leak from the pump seals due to excessive movement in the shaft bearings, sometimes caused by over-tightening of drive belts. Coolant usually leaks from a relief hole below the pump when the seals are faulty. Always check for leaks in this area.

Cooling fans

There are three types of cooling fan are used in automotive cooling systems.

- *Electric fans* are the most common type used by vehicle manufacturers. Most engines on front-wheel-drive vehicles are

mounted in an east–west configuration, making it impossible for the engine crankshaft to drive a cooling fan. Driving a cooling fan also takes valuable engine power, while an electric fan can be connected to a temperature sensor through the engine computer that switches on the fan only when needed, e.g. in traffic or when the air-conditioning pressure rises above a certain pressure (1400 kPa). Some manufacturers connect the fan so that it will operate when the air conditioning is switched on or engine temperature rises above a set point. Others use separate electric fans: one for the cooling system and one or both for the air conditioning. Note that these can start at any time that the engine temperature rises above the set sensor reading, even in some cases when the engine is turned off.
› *Mechanical drive fans* operate continuously and are fixed directly to the water pump shaft.
› *Viscous coupled fans* are also engine-driven. They incorporate a finned housing between the pump shaft and the fan that contains a temperature-sensitive silicon fluid. As air is blown from the radiator, it moves over the finned section of the fan housing and becomes hotter, solidifying the silicon fluid, which now starts to transmit drive to the fan. As the radiator air cools down, the silicon changes back to a liquid and drive is reduced. This saves valuable engine power and reduces fan noise.

Coolant

Water is able to hold huge amounts of heat (it has a high specific-heat value) so it is therefore the main ingredient in a coolant mixture. However, a coolant must also contain additives that will:

› lower the temperature at which the coolant freezes (anti-freeze or ethylene-glycol)
› increase the temperature at which the coolant boils (anti-boil additives)
› prevent corrosion due to electrolysis (corrosion inhibitor).

The anti-freeze capabilities of the coolant can be increased by adding a greater concentration of ethylene-glycol to water. This should be done if very cold conditions are likely. If the coolant freezes, the ice will expand, cracking the cylinder block.

Coolants should be changed every two to four years on average. However, always check in the vehicle handbook for the manufacturer's recommendations. Coolant concentrations are usually between 33 and 50 per cent of the total coolant system capacity. Use the correct coolant and concentration specified for the vehicle being serviced. Testing of the coolant can be carried out by using coolant strips, a special tool called a coolant hydrometer, or a spectrometer.

Heater and air-conditioning case/ducting

The ducting of a heating and air-conditioning system distributes both heated and cooled air to various vent outlets. It also houses the heater and evaporator cores and controls where the air comes from.

It consists of three sections (see Figure 13.1):

› air intake (fresh, re-circulate)
› core section (evaporator, heater)
› air distribution (face, feet, demist and so on).

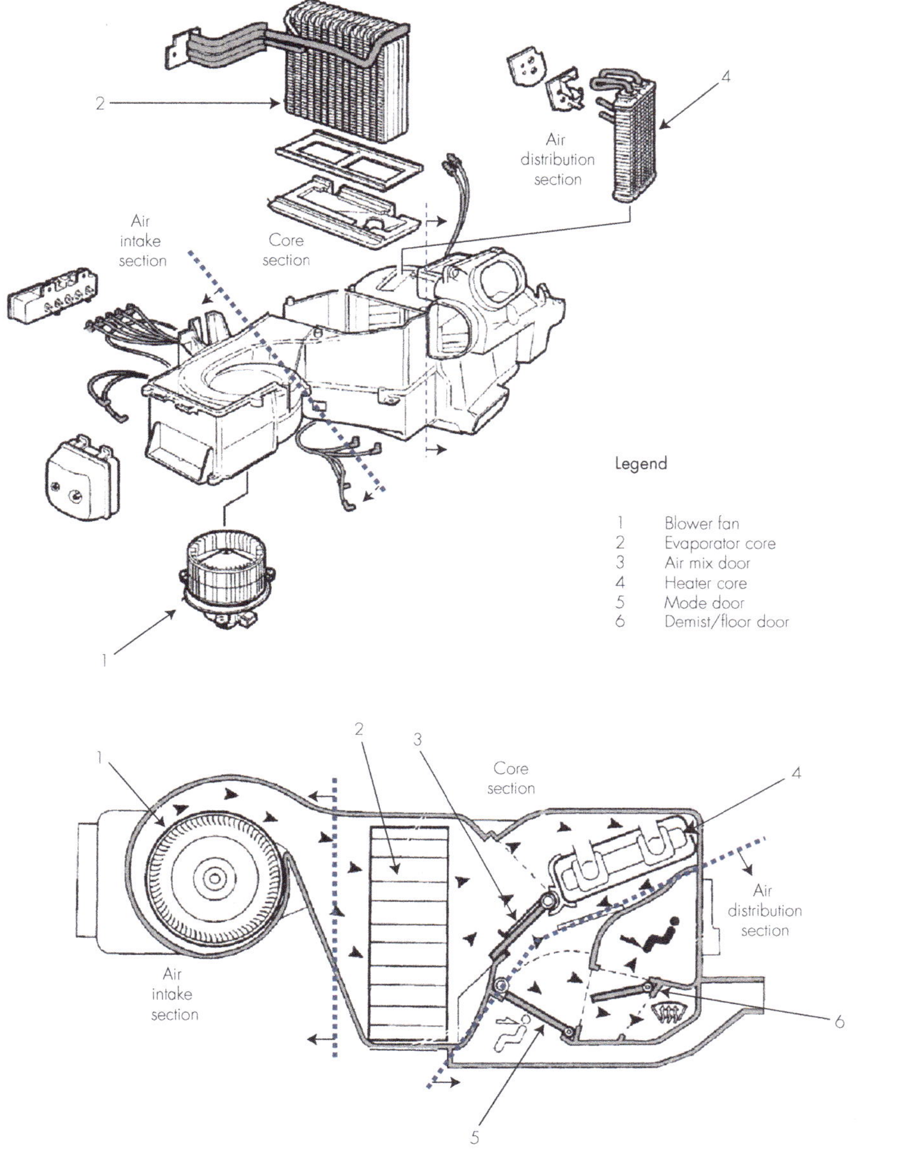

Figure 13.1 Components of the cooling, heating and air-conditioning systems
Source: General Motors Holden

Air intake

The air intake or inlet section of the air-conditioning case consists of two inlet vents (one that allows fresh air in and one taking air from inside the vehicle), a vent door, a vent actuator motor, and a blower fan and a motor (see Figure 13.1).

The vent door or flap is controlled by the driver operating the re-circulate/fresh control lever on the dashboard. With the lever in the 'fresh' position, air is taken from the outside air chamber, situated under the wiper arms on most vehicles. If the lever is moved to the re-circulate position ('recirc'), air is taken from inside the vehicle cabin. This makes the unit operate more effectively, as the air that has been heated or cooled by the system is then reheated or re-cooled. The re-circulating air mode is also useful when driving through areas that have unpleasant odours, or when following polluting vehicles. Note 're-circulate' should not be used for extensive periods of time as the oxygen level inside the cabin will decrease and the occupants can become drowsy.

Most vehicles now incorporate a pollen filter that is used to filter the air as it moves through the intake section. This filter should be replaced at manufacturer's specifications. Access is obtained behind the glove box on most systems.

Core section

The core section (sometimes called the plenum section) connects directly to the intake section of the ducting and is the centre part of the case assembly. It houses the **heater core**, the evaporator core, the **blend doors** and actuator motors (see Figure 13.1).

The evaporator core is located in such a position that all the air will pass through it, whether the air conditioning is switched on or not. The air can then either enter or bypass the heater core by moving the blend door. The blend doors (or 'mixing doors') are moved by cable – or more commonly vacuum or electric motor. Changing the position of the temperature control on the dashboard will alter the temperature of the interior of the vehicle, as it alters the percentages of hot and cold air blended together.

Heating

The heater core is similar to a small radiator and is heated by the engine coolant that flows through it. Coolant flow is sometimes controlled by a heater valve that is opened when the temperature control is raised. However, most manufacturers now leave the coolant flowing through the heater at all times, and use a blend door to control the temperature by controlling the air flow through the heater. This prevents problems with the heater valves, but the blend door flap can cause problems if it becomes faulty.

When the temperature control is moved to the hot position, the blend door opens fully to force all the air from the intake section through the evaporator and then through the heater core. The driver now has the choice of either hot air or, if the air conditioning is switched on, heated/air-conditioned air. This is moisture-reduced to help demist the interior of the vehicle.

Cooling

When the driver selects 'maximum cool', all the air from the evaporator bypasses the heater core. The blend door prevents air from flowing through the heater core. Air flows from the intake section through the evaporator to the distribution section of the air-conditioning/duct case.

If the driver requires slightly less cooling, the blend door can be opened slightly to allow a small amount of air-conditioning air to move through the heater and on to the vents.

Some vehicles use two or more blend doors to select different temperatures at different sections of the car through their individual vents.

It is important that blend doors are adjusted correctly. If the air-conditioning system is switched to maximum cool and the blend door is slightly open when it should be closed, the air-conditioning system may be diagnosed as being inefficient. To check this, switch the air conditioning off, put the blower fan on low and set the temperature control at the cold position. Ambient temperature air should now be flowing from the vents. Simply check the ambient air temperature and the vent temperature, and compare the readings. If the vent temperature is markedly higher than the ambient temperature, the heater is influencing this difference in temperature.

Air distribution

The air distribution section consists of a series of doors or flaps that direct air to: the windscreen for demisting purposes; the face vents or the floor vents; or a combination for heating or cooling (see Figure 13.1).

Some vehicles use a bi-level mode whereby the floor vent has warm air delivered to the feet, while cool air is directed at the face or demist to keep the driver alert or to keep the screen clear.

For efficient diagnosis, it is important to be familiar with the type of system being worked on.

NOTE

+ Always consult the manufacturer's manual. The owner's manual can also be useful in explaining correct system operation.

Chapter-revision questions

PRACTICAL EXERCISES

1. Pressure test the cooling system and note any leakages or drops in pressure.
2. Pressure test the radiator cap and note the release pressure. Compare this with the manufacturer's specifications.
3. Test the cooling system thermostat.
   - Does the engine warm up quickly?
   - Does it maintain a constant engine temperature?
4. Test the vent door and blower fan operation.
   - Note any abnormal fan noises.
   - Check that vents close and open when selected.
5. Check the blend door and heater valve operation (if fitted).
   - Check the ambient air temperature and the vent temperature (with the air conditioning switched off) to ensure there is only a small difference in the readings.

SELF-CHECK QUESTIONS

1. What are the two types of radiators?
2. What are the two functions of a radiator cap?
3. Why is a cooling system pressurised?
4. What function does a vacuum valve perform in a radiator cap?
5. What is the function of the thermostat?
6. Why are manufacturers fitting thermostats into the lower hose?
7. What are the three ways a water pump may be driven?
8. What are the three types of cooling fans found on modern vehicles?
9. What are the three functions of coolant?
10. What are the three sections of the ducting assembly?

CHAPTER 14

Electronic climate control (ECC)

Objectives

On completion of this chapter the student will be able to:

- state the purpose of the ECC
- identify each component of the ECC and state its purpose.

This chapter covers some of the essential knowledge and associated skills involved with:

- AURT222670A Servicing air-conditioning systems
- AURT366108A Carry out diagnostic procedures.

In this system both the heating and cooling functions are controlled by a single control module unit.

The passenger compartment environment can be maintained at a pre-set comfort level, with the driver controlling the heating, cooling, **dehumidifying** and venting operations by selecting positions on the control module.

Automatic temperature control (ATC) is more commonly known as the electronic climate control (**ECC**) system. It incorporates a central control computer and several sensors to monitor the complete system operation. These include the following temperature sensors:

- cabin
- ambient
- engine coolant
- evaporator.

In addition there are sensors for:

- **sun load**
- blower speed
- selected temperature setting
- blend door position.

Using the information that they have gathered, these sensors control the operation of the appropriate air ducts (face, feet, demist or any combination) and provide the heating, cooling and venting necessary for achieving the selected comfort temperature in the least possible time. They must then maintain this temperature under differing conditions. The ECC also controls the relative humidity within the passenger area, keeping it at about 48 per cent. This also prevents windows from fogging during periods of high humidity. The comfort setting can be set at any temperature level between 16 and 30°C.

For example, if the ECC module is set at 25°C, the environment inside the compartment will be maintained at 25°C and the relative humidity will be held at between 46 and 50 per cent.

On a day when the outside temperature is 45°C, a correctly operating ECC system can rapidly cool the interior of the vehicle to a predetermined temperature and then adjust itself to maintain the desired temperature level. On days when the outside temperature is 10°C, the system rapidly heats the vehicle's interior and blends the heater and evaporator air to maintain the desired temperature. In current model vehicles comfort level can be set to the individual's preferred setting; i.e. the passenger and driver can each set their own temperature – as each is controlled separately.

Another recent innovation is the introduction of infrared dual-zone climate control systems that read the body temperature of each front occupant, and adjusts their individual temperature settings accordingly.

A number of different ECC systems are available and it is therefore impossible to describe every system in detail. Here we cover the operation of the main components found in these systems. Manufacturers' workshop manuals must be consulted whenever a diagnostic procedure is required.

NOTE

+ A digital ohm meter with a 20 MΩ (mega-ohm) impedance must be used whenever the manufacturer recommends that a component's resistance be measured. Failure to do so can result in component and/or computer damage.

Most, if not all, ECC systems have self-diagnostic facilities that enable the technician to trace faults. The processor will tell them where the fault is. However, the display code only indicates three things:

› open circuit
› short circuit
› sensor out of the range of normal resistance values.

In most ECC systems, if it is necessary to carry out a performance test, the unit should be placed in manual mode and manufacturer's procedures followed (or the steps outlined in Service Procedure 2, Performance testing, on page 189).

Component parts

The electronic control unit (ECU)

This unit receives information from the system sensors marked as inputs on Figure 14.1. It then powers components to operate or move, to control system operation and temperature. These are called outputs in Figure 14.1.

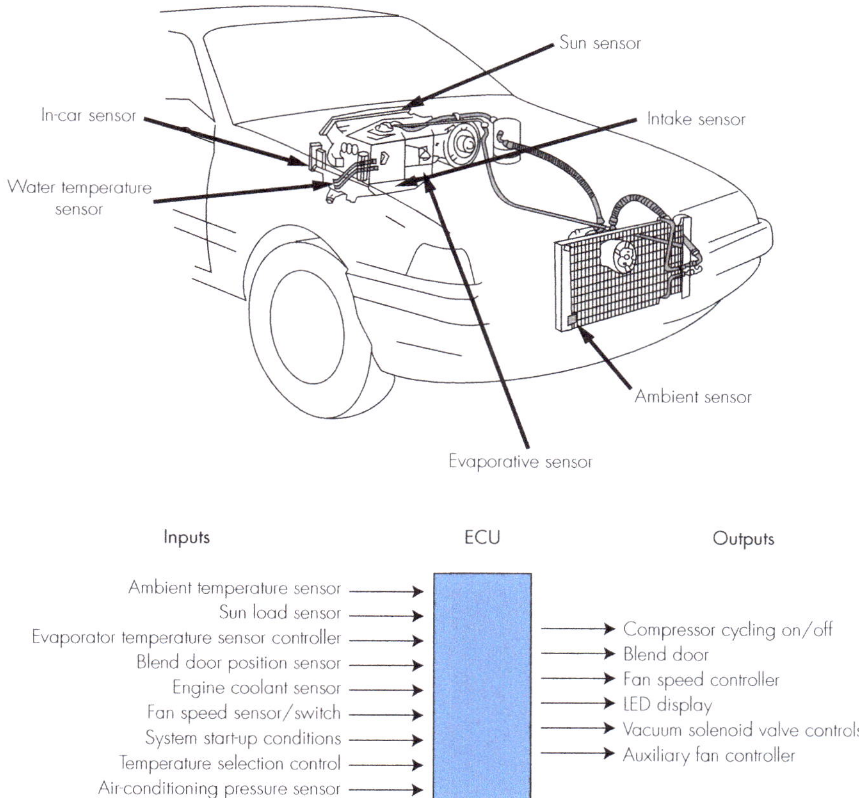

Figure 14.1 Sensor positions in the air-conditioning system and ECU outputs in response to inputs from the sensors

The ECU performs the following tasks:

- When the engine is running, the control unit remembers the pre-set temperature.
- It reads the inputs from all sensors.
- It sets the position and speeds of the outputs so as to bring the interior temperature of the vehicle to the pre-set level.
- It provides power through the electrical circuit of the system to operate the compressor.
- It monitors all sensors to continuously check and adjust the system controls and so maintains the pre-set temperature.

Ambient air sensor

The ambient air sensor is usually located just behind the grill, in front of the condenser. It monitors the air temperature before it is heated, as it passes through the condenser and radiator (see Figure 14.2). This information is then passed on to the control unit.

This sensor is a thermistor, i.e. it converts a temperature into a resistance. Therefore, as the temperature changes, the resistance will also change (see Figure 14.3). This information is then used by the ECU to help with its control of the system.

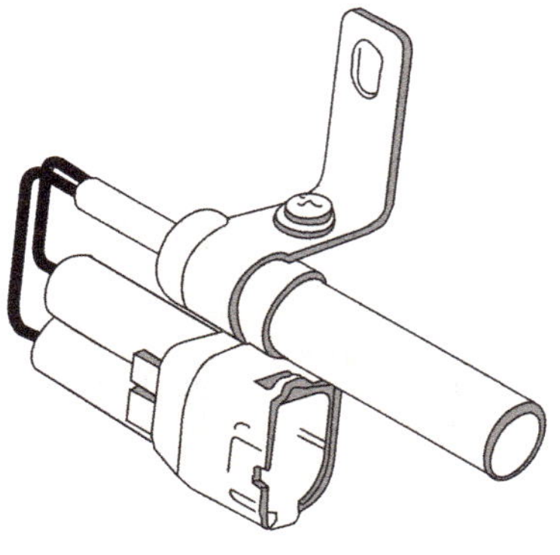

Figure 14.2 An ambient temperature sensor
Source: Ford Motor Co.

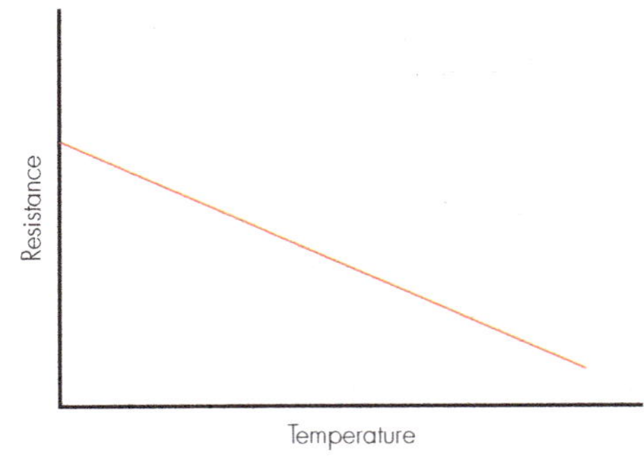

Figure 14.3 Graph demonstrating how resistance changes with temperature

Sun load sensor

The **sun load sensor** is also known as a 'photo sensor' or 'solar sensor'. It is usually placed in the centre of the dash behind the windscreen, next to the demister vents (see Figure 14.4). The sun load sensor is a photo-voltaic diode that sends a voltage signal to the ECU when strong sunlight falls on it. This tells the ECU to provide extra cooling when the vehicle is facing direct sunlight by increasing the fan speed and changing cycling times or (in the case of the variable stroke compressor) compressor capacity.

In-car temperature sensor

This sensor is also a thermistor, and is usually mounted near the control unit or in the lower part of the dash. It is also known as an 'aspirator' (see Figure 14.5).

A small amount of air is drawn into the sensor by **venturi action** and across the thermistor. This tells the control unit what the temperature of the air inside the vehicle is.

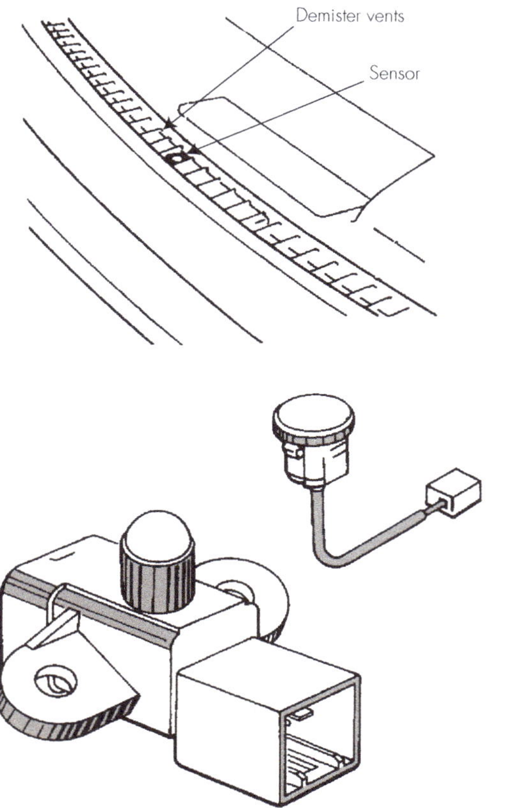

Figure 14.4 Typical sun load sensors (below) and placed in position on the demister grill (above)

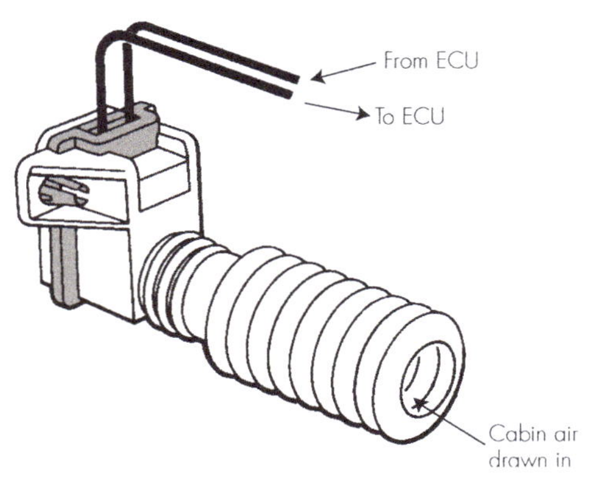

Figure 14.5 An in-car temperature sensor

Evaporator temperature sensors

Electronic climate control systems have either one or two sensors in the evaporator. They are both thermistors (see Figure 14.6).

- *Intake air sensor* – this measures the temperature of the air before it passes into the evaporator. This can either take the place of, or be complementary to, the ambient air sensor.
- *Evaporator sensor* – attached to the evaporator air outlet, and measures the temperature of the air as it leaves the evaporator. This information will either cycle the compressor or reduce its capacity (variable stroke compressors).

Engine coolant temperature sensor

This sensor is a thermistor that provides engine coolant temperature information to the ECU processor. On the initial engine start-up, this sensor may delay the operation of the ECC system until the engine temperature has reached a predetermined setting. If the engine coolant temperature rises above its normal setting, it can stop the operation of the system to help the engine to cool down.

The sensor is found usually near the other engine coolant sensors.

Fan speed controller

In manual mode, the fan speed switch tells the ECC module which speed is to be selected. In auto mode the fan speed is selected by the ECC module in response to system needs. In auto mode

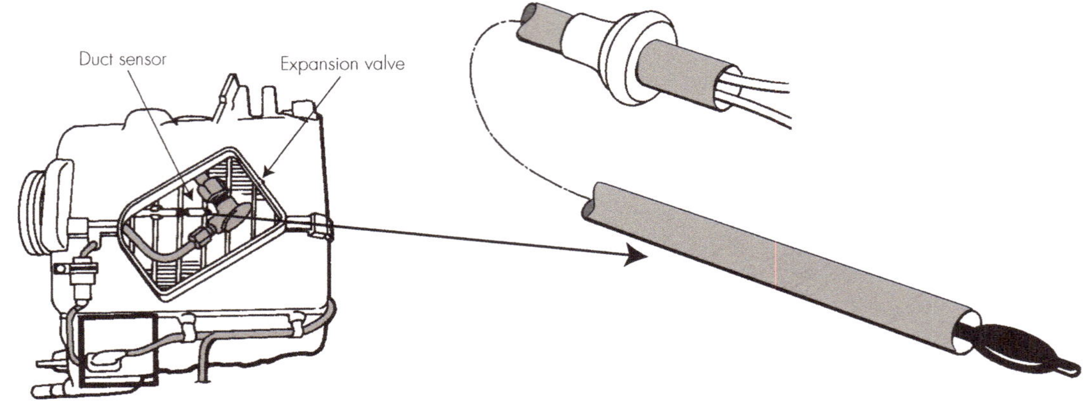

Figure 14.6 Position of the thermistor in the evaporator sensor (left) and an enlargement of the thermistor (right)

there can be up to 13 different speed selections, but there are only three to five speeds available in manual mode.

Blend door or air mix door actuators

These are usually electric motors that accept electronic commands from the ECC module and then position the air mix doors or blend doors in order to mix cold air from the evaporator and hot air from the heater core. In this way they help to keep the interior environment at the pre-set temperature. These doors are at the heart of the electronic climate control system (see Figure 14.7).

When the system is working in the auto mode, these doors remain in one position until the desired temperature has been reached. They then constantly change their angle, varying the mix of cold and hot air and so maintaining the pre-set environment. In a dual-zoned system, where the passenger and driver can have different comfort settings, there is more than one air mix door.

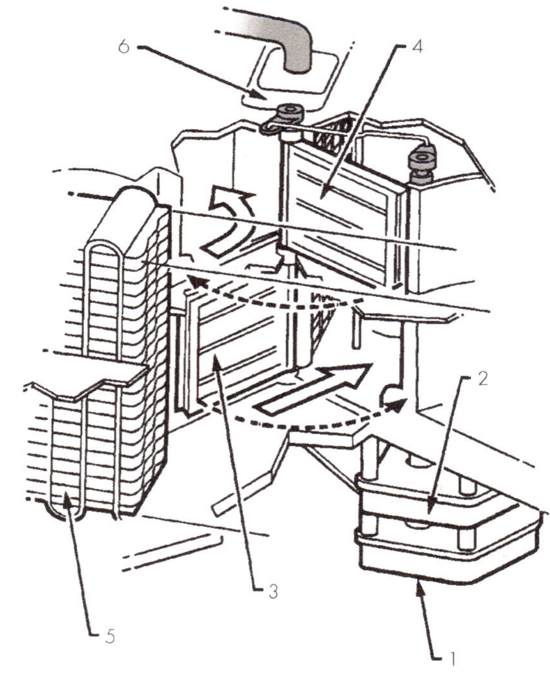

1 Primary air mix motor (driver's side)
2 Secondary air mix motor (passenger side)
3 Air mix door lower (driver's side)
4 Air mix door upper (passenger side)
5 Evaporator core
6 Heater core

Figure 14.7 Operation of the air mix doors

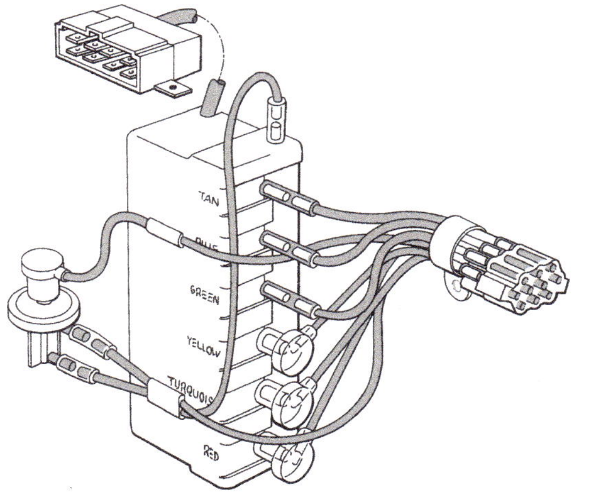

Figure 14.8 A vacuum solenoid pack

Vacuum solenoid pack

The function of the vacuum solenoid pack unit is to distribute vacuum from the reserve tank and engine to the various flap actuators and heater control tap (see Figure 14.8).

The system consists of electrical solenoids that control the vacuum distribution to different **vacuum servo units**. The opening and closing of these servos control the air movement to the face, feet, demist or any combination.

Double-vacuum servo units allow the flaps that they control to be closed, open or set halfway, and this helps to blend the air more accurately (see Figure 14.9).

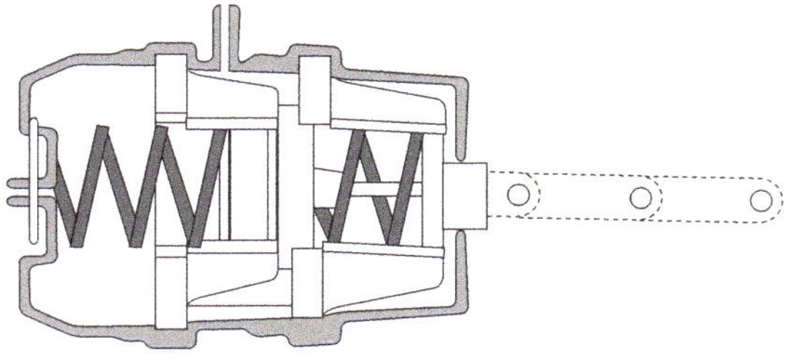

Figure 14.9 A double-vacuum servo unit

Blower speed controller or amplifier

This controller is located in the air plenum chamber, near the system's fan unit. The controller converts a low-current signal from the ECC module to a high-current, variable voltage signal that operates the fan motor at its different speeds. A delay function in auto mode allows for a gradual increase or decrease in fan speed under all conditions.

Temperature selection switch

The two types of switch currently in use (see Figure 14.10) are:

- a temperature switch that is a variable resistor. This informs the ECC about the desired temperature setting
- a LED module and signal amplifier with touch-button operation that can either raise or lower the desired temperature setting.

NOTE

+ The compressor surge diode is an important protective device because of the large number of electronic devices in the ECC system.

The collapse of the magnetic field as the clutch coil is released can cause high spike voltages (EMF) that are harmful to all electronic components. The diode is therefore placed across the clutch coil to provide a path to ground, and so keep these spike voltages at a safe level.

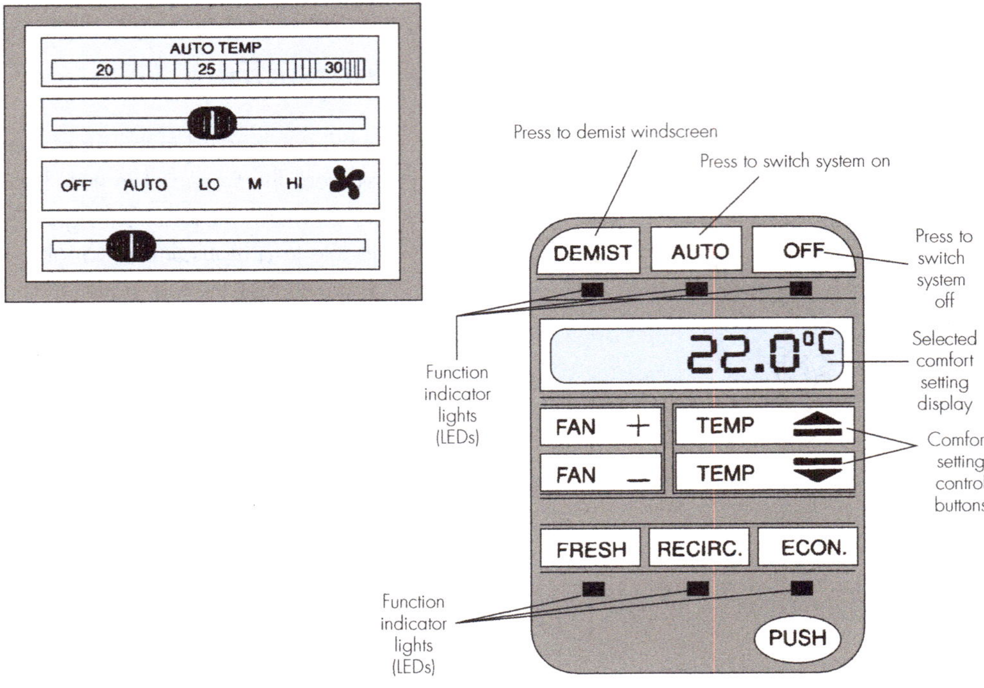

Figure 14.10 A temperature selection switch (left) and an LED with touch-button control (right)

Using scan tools to access ECU air-conditioning data

Modern scan tools can be used to access data from vehicle electronic control units (ECU). This will allow technicians to:

› view fault codes that are stored in the ECU's memory
› view live data concerning the air-conditioning system
› view graphs of air-conditioning system performance.

Scan tools are connected to the vehicle ECU via the diagnostic link connector (DLC), which is usually located under the dash on the driver's side (see Figure 14.11).

Fault codes

If a fault occurs in the air-conditioning system it will log a code in the ECU's RAM memory. This code only indicates that a component is outside its normal parameters, i.e. lower or higher voltage or open circuit. These codes will allow the technician to make a diagnosis as to the cause of the problem but the code may only be an indication as to the area of the fault and all related components should be tested for correct operation before replacement. Figure 14.12 shows a screen dump of typical fault code readout on a modern scan tool.

Live data streaming

Another feature of scan tools is the ability to live stream data from the vehicle's air-conditioning

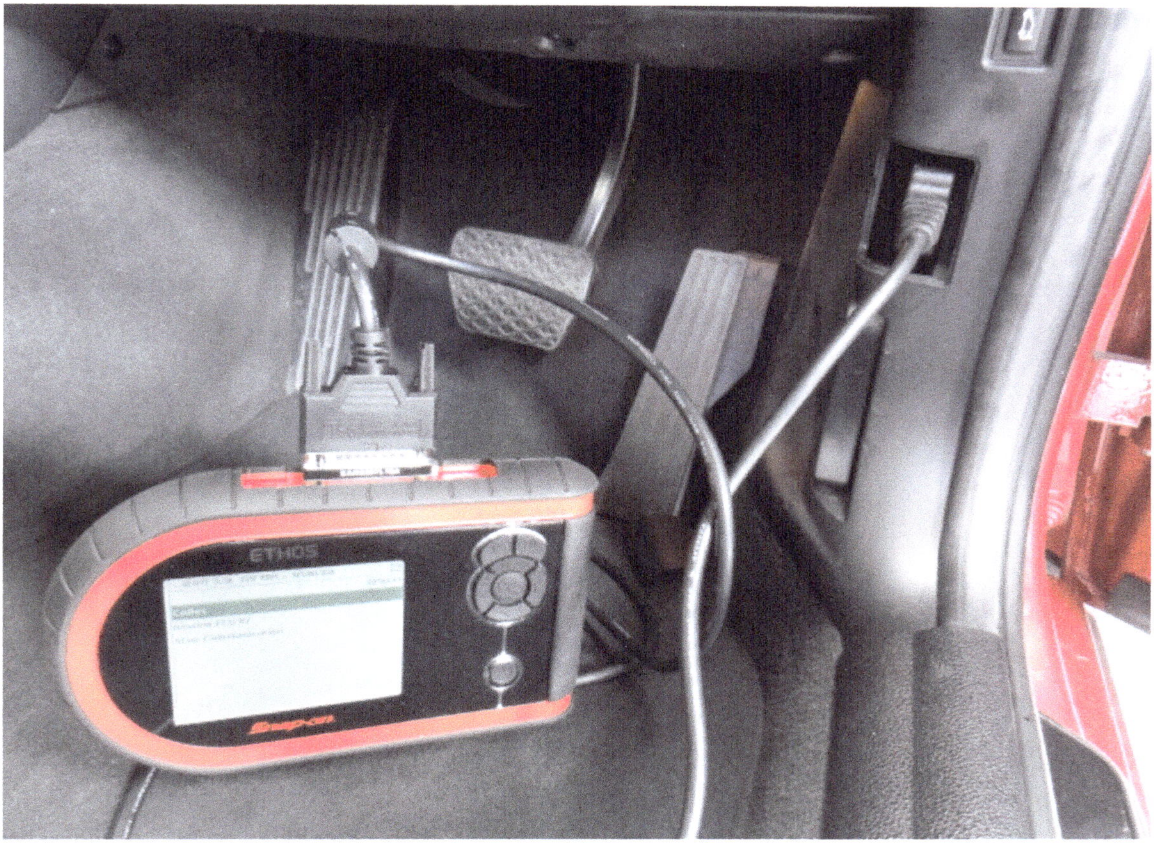

Figure 14.11 Connecting a scan tool to the vehicle

Figure 14.12 Screen capture of fault codes using scan tool
Source: Reproduced with permission of CODA

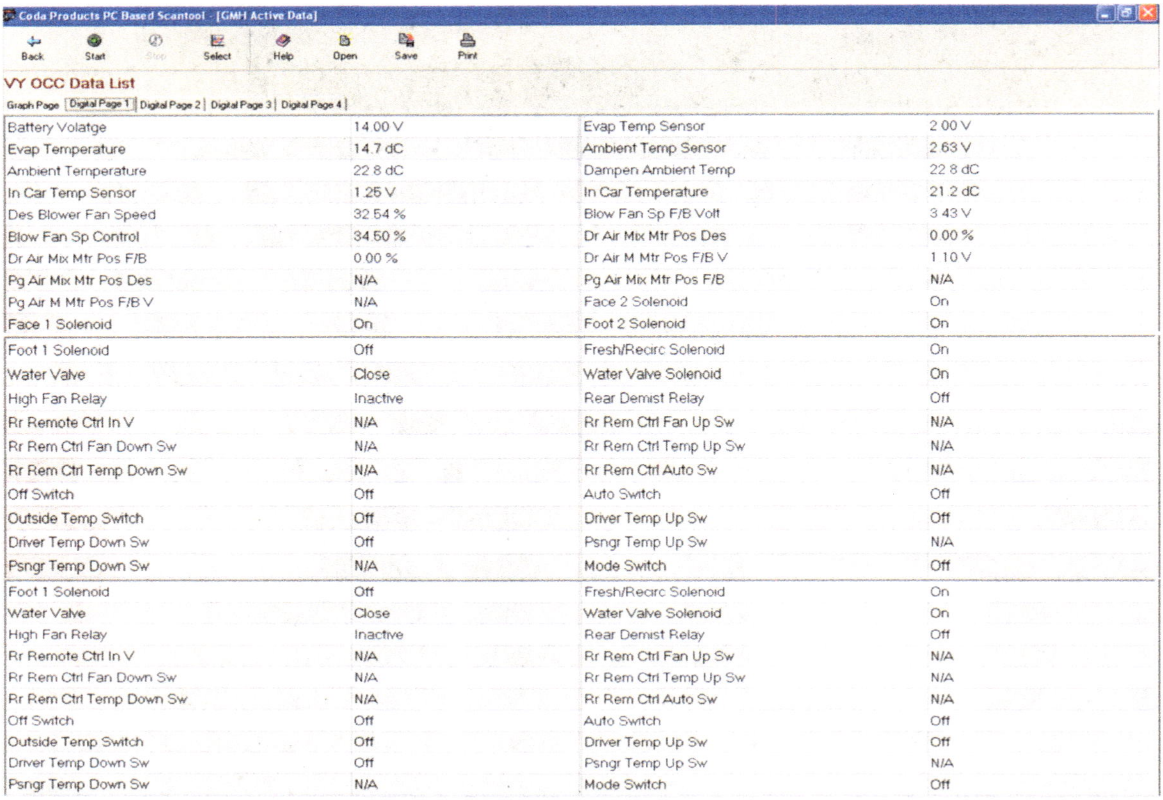

Coda Products PC Based Scantool - [GMH Active Data]

Back Start Stop Select Help Open Save Print

VY OCC Data List

Graph Page | Digital Page 1 | Digital Page 2 | Digital Page 3 | Digital Page 4

Parameter	Value	Parameter	Value
Battery Volatge	14.00 V	Evap Temp Sensor	2.00 V
Evap Temperature	14.7 dC	Ambient Temp Sensor	2.63 V
Ambient Temperature	22.8 dC	Dampen Ambient Temp	22.8 dC
In Car Temp Sensor	1.25 V	In Car Temperature	21.2 dC
Des Blower Fan Speed	32.54 %	Blow Fan Sp F/B Volt	3.43 V
Blow Fan Sp Control	34.50 %	Dr Air Mix Mtr Pos Des	0.00 %
Dr Air Mix Mtr Pos F/B	0.00 %	Dr Air M Mtr Pos F/B V	1.10 V
Pg Air Mix Mtr Pos Des	N/A	Pg Air Mix Mtr Pos F/B	N/A
Pg Air M Mtr Pos F/B V	N/A	Face 2 Solenoid	On
Face 1 Solenoid	On	Foot 2 Solenoid	On
Foot 1 Solenoid	Off	Fresh/Recirc Solenoid	On
Water Valve	Close	Water Valve Solenoid	On
High Fan Relay	Inactive	Rear Demist Relay	Off
Rr Remote Ctrl In V	N/A	Rr Rem Ctrl Fan Up Sw	N/A
Rr Rem Ctrl Fan Down Sw	N/A	Rr Rem Ctrl Temp Up Sw	N/A
Rr Rem Ctrl Temp Down Sw	N/A	Rr Rem Ctrl Auto Sw	N/A
Off Switch	Off	Auto Switch	Off
Outside Temp Switch	Off	Driver Temp Up Sw	Off
Driver Temp Down Sw	Off	Psngr Temp Up Sw	N/A
Psngr Temp Down Sw	N/A	Mode Switch	Off
Foot 1 Solenoid	Off	Fresh/Recirc Solenoid	On
Water Valve	Close	Water Valve Solenoid	On
High Fan Relay	Inactive	Rear Demist Relay	Off
Rr Remote Ctrl In V	N/A	Rr Rem Ctrl Fan Up Sw	N/A
Rr Rem Ctrl Fan Down Sw	N/A	Rr Rem Ctrl Temp Up Sw	N/A
Rr Rem Ctrl Temp Down Sw	N/A	Rr Rem Ctrl Auto Sw	N/A
Off Switch	Off	Auto Switch	Off
Outside Temp Switch	Off	Driver Temp Up Sw	Off
Driver Temp Down Sw	Off	Psngr Temp Up Sw	N/A
Psngr Temp Down Sw	N/A	Mode Switch	Off

Figure 14.13 Screen capture of live data
Source: Reproduced with permission of CODA

system while it is in operation, this allows the technician to view system component parameters at the same time and diagnose faults or anomalies with system operation. See Figure 14.13.

System graphs

Scan tools can also allow the selection of several system outputs to be seen on a system graph.

This allows system performance to be evaluated by overlaying outputs such as compressor on/off, and/or high- and low-side temperatures on a screen together. Figure 14.14 shows a dotted line running down the left side of the screen and the relevant data in the boxes on the left. Figures 14.15 and 14.16 show the line has now been moved to the right along the screen and the changes in data as the system stabilises.

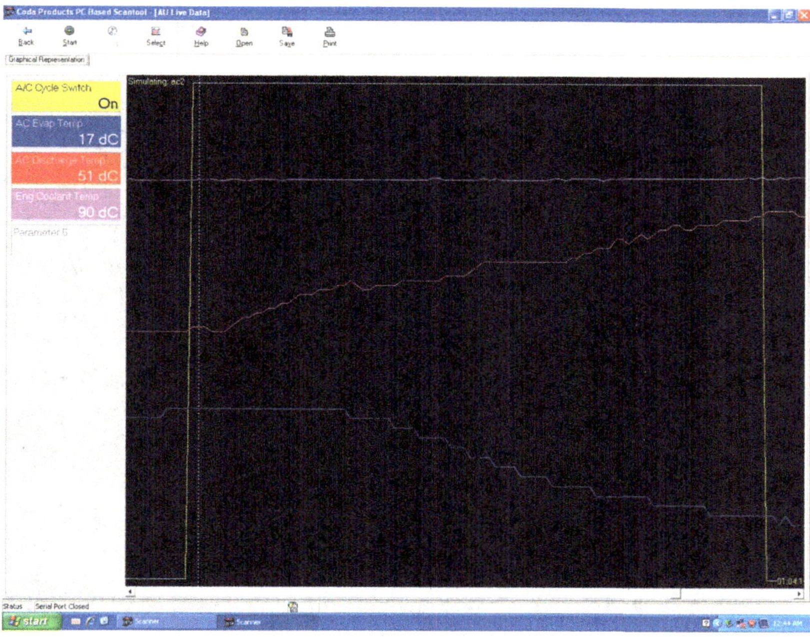

Figure 14.14 Screen capture of air-conditioning outputs using scan tool (1)
Source: Reproduced with permission of CODA

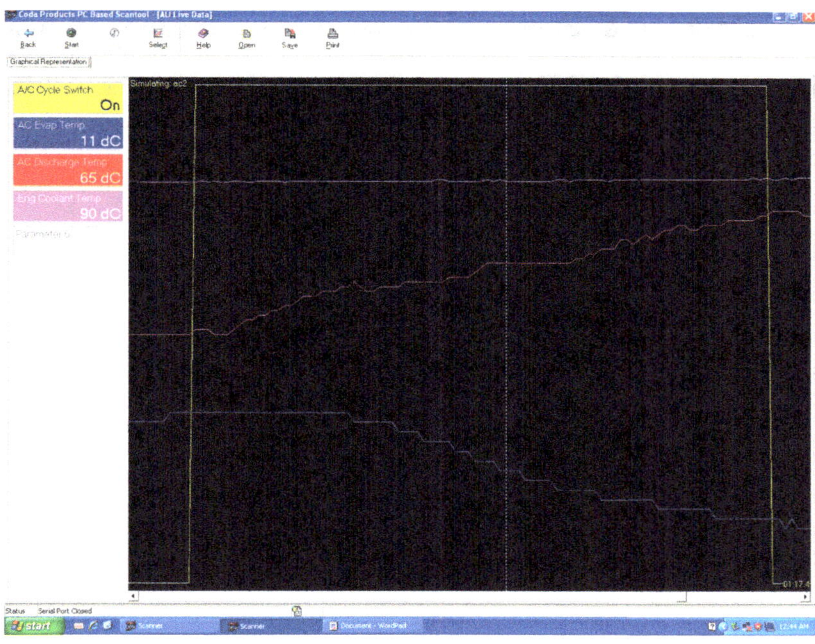

Figure 14.15 Screen capture of air-conditioning outputs using scan tool (2)
Source: Reproduced with permission of CODA

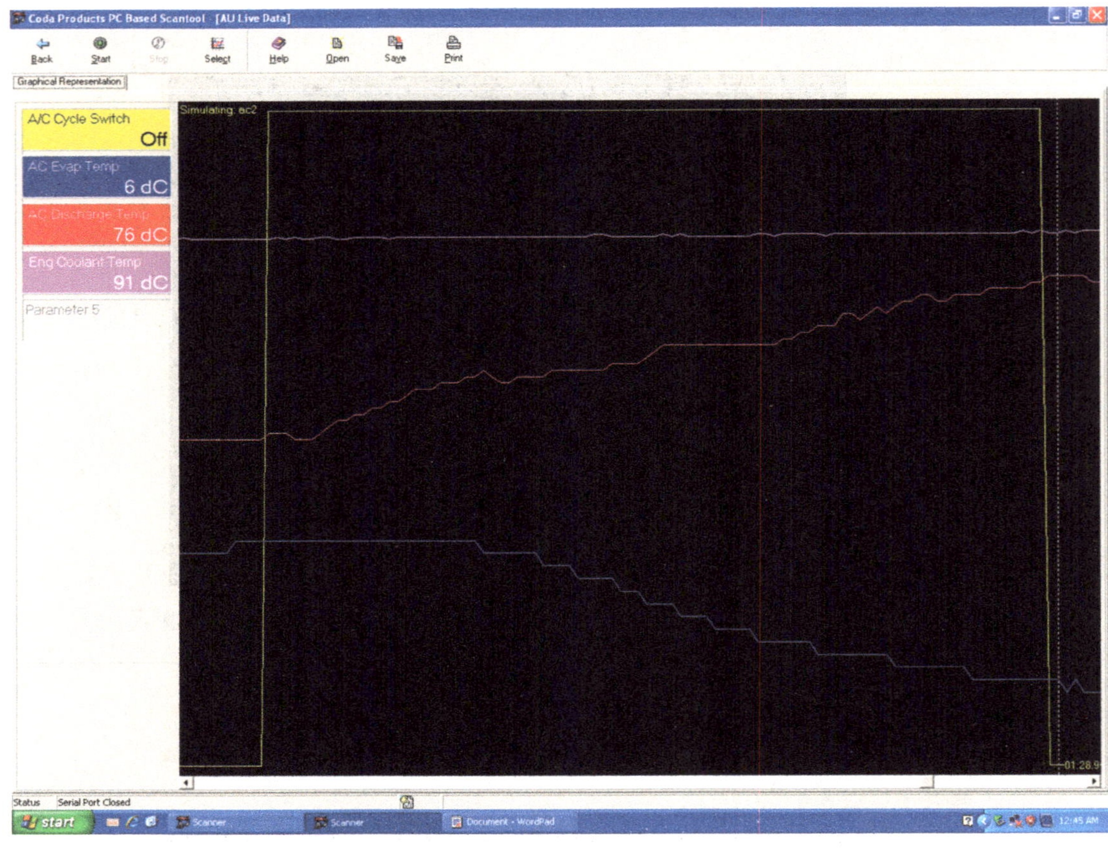

Figure 14.16 Screen capture of air-conditioning outputs using scan tool (3)
Source: Reproduced with permission of CODA

Chapter-revision questions

PRACTICAL EXERCISE

1 Connect a scan tool to a late-model vehicle with climate control and assess the following:
- Check to see if any fault codes have been logged, if there are none present disconnect the ambient temperature sensor and re-scan for codes.
- Remove fault codes using the scan tool.
- Live stream the air-conditioning system data and observe evaporator and condenser temperatures.

SELF-CHECK QUESTIONS

1 What does the term electronic climate control mean?
2 What are the main sensors in climate-controlled systems?
3 What humidity level does the ECC maintain?
4 What instrument should be used when testing sensor function?
5 Where is the ambient air sensor usually located?
6 What type of diode is a sun sensor?
7 Will a thermistor increase or decrease its resistance as it gets warmer?
8 What is the function of the blend door?
9 What is used to prevent voltage spikes damaging the ECU and other electronic components?
10 What functions can a scan tool perform?

Section 2
System diagnosis

Introduction

Service technicians must understand the function of the components if they are to be able to diagnose the air-conditioning system. They must also be able to interpret the high- and low-pressure gauge readings that they have obtained.

The use of the refrigeration manifold and gauges can be compared to the use of a doctor's stethoscope.

An abnormal gauge reading will always relate to a specific problem or group of associated problems. A low-pressure gauge reading shows the temperature of the liquid refrigerant as it vaporises while passing through the evaporator and the operation of the thermal expansion valve, while the high-pressure gauge reading shows the temperature of the high-pressure vapour as it condenses, passing through the condenser.

Any deviation from the normal pressure reading indicates that there is a system malfunction caused by:

- a faulty control device
- a restriction
- a defective component
- an overheated engine.

Any of these malfunctions will affect the performance of the whole system.

The pressure as shown on the gauges can be converted to a temperature. The temperature scale is shown on the inner ring. Therefore, a pressure reading indicates the temperature of the refrigerant.

Low-pressure/temperature = low-pressure refrigerant vaporising pressure/temperature

High-pressure/temperature = high-pressure refrigerant condensing pressure/temperature.

This section includes:

- system-diagnosis charts and flow charts
- common faults charts
- refrigerant pressure/temperature charts
- system operating pressure/temperature charts
- pressure/temperature exercises
- system diagnostic exercises.

2.1

System-diagnosis flow charts

System-diagnosis flow chart 1

Check these five points

1 Test gauges connected
2 All gauge hoses purged
3 System stabilised
4 Performance test made
5 Gauge readings noted

Reasons for no cooling from system

1 Blown fuse
2 Broken or disconnected electrical wire
3 Broken or disconnected ground wire
4 Clutch coil or solenoid burnt out or disconnected
5 Electric switch contacts in thermostat burned excessively, or sensing element defective
6 Blower motor disconnected or burnt out
7 Ignition switch ground or relay burnt out
8 Loose or broken drive belt
9 Compressor partially or completely frozen
10 Compressor reed valves inoperative – indicated by slight variation of both gauge readings at any engine speed
11 Expansion valve stuck open – indicated by normal discharge pressure, high suction pressure and evaporator flooding
12 Heater valve inoperative – indicated by hot water in heater and hot discharge air from evaporator
13 Broken refrigerant line
14 Fusible plug blown (not used on all systems)
15 Leak in system
16 Clogged screen or screens in receiver dryer or expansion valve; plugged hose or coil
17 Compressor shaft seal leaking

Reasons for intermittent cooling

1. Defective circuit breaker, blower switch or blower motor
2. Bad earth connection or loose electrical connection in compressor clutch coil or solenoid
3. Compressor clutch slipping
4. Expansion valve icing up – may be caused by excessive moisture in the system or incorrect superheat adjustment
5. Evaporator coil icing up – thermostat probe not in coil fins, thermostat adjusted too low, defective thermostat
6. Clogged evaporator fins

Reasons for insufficient cooling from system

1. Blower motor sluggish
2. Compressor clutch slipping
3. Obstructed blower discharge passage
4. Clogged air intake filter
5. Insufficient air circulation over condenser coils (fins clogged with dirt or bugs)
6. Evaporator clogged
7. Outside air vents open
8. Insufficient refrigerant in system
9. Clogged screen in expansion valve – indicated by gauge pressures being normal or showing slightly increased discharge pressure and low suction pressure with evaporator air output temperature high
10. Expansion valve thermal bulb has lost its charge – indicated by too high a low-gauge reading and excessive sweating of evaporator and suction line
11. Clogged screen in receiver – indicated by higher than normal reading on high-pressure gauge, lower than normal reading on low-pressure gauge, and liquid lines cold to touch, with possible frost
12. Excessive moisture in system – indicated by excessive head-pressure gauge reading
13. Air in system – indicated by excessive head pressure
14. Thermostat defective or improperly adjusted – indicated by low-gauge reading high or clutch cycling at too high a reading
15. Vehicle heater partially operating

Reasons for noisy system

1. Defective winding or improper connection in compressor clutch coil or solenoid
2. Loose or excessively worn drive belts
3. Noisy clutch
4. Compressor noisy – loose mounting or worn inner parts
5. Loose panels on car
6. Compressor oil level low
7. Blower fan noisy – excessive wear in motor
8. Idler pulley and bearing defective
9. Excessive charge in system – rumbling noise or vibration in high-pressure line, thumping noise in compressor, excessive head pressure and suction pressure, bubbles or cloudiness in sight glass, or low head pressure
10. Low charge in system – hissing in evaporator case at expansion valve or low head pressure
11. Excessive moisture in system – expansion valve noisy, suction pressure low

Source: Sanden International (Australia) Pty Ltd

System-diagnosis flow chart 2 – the Sanden system

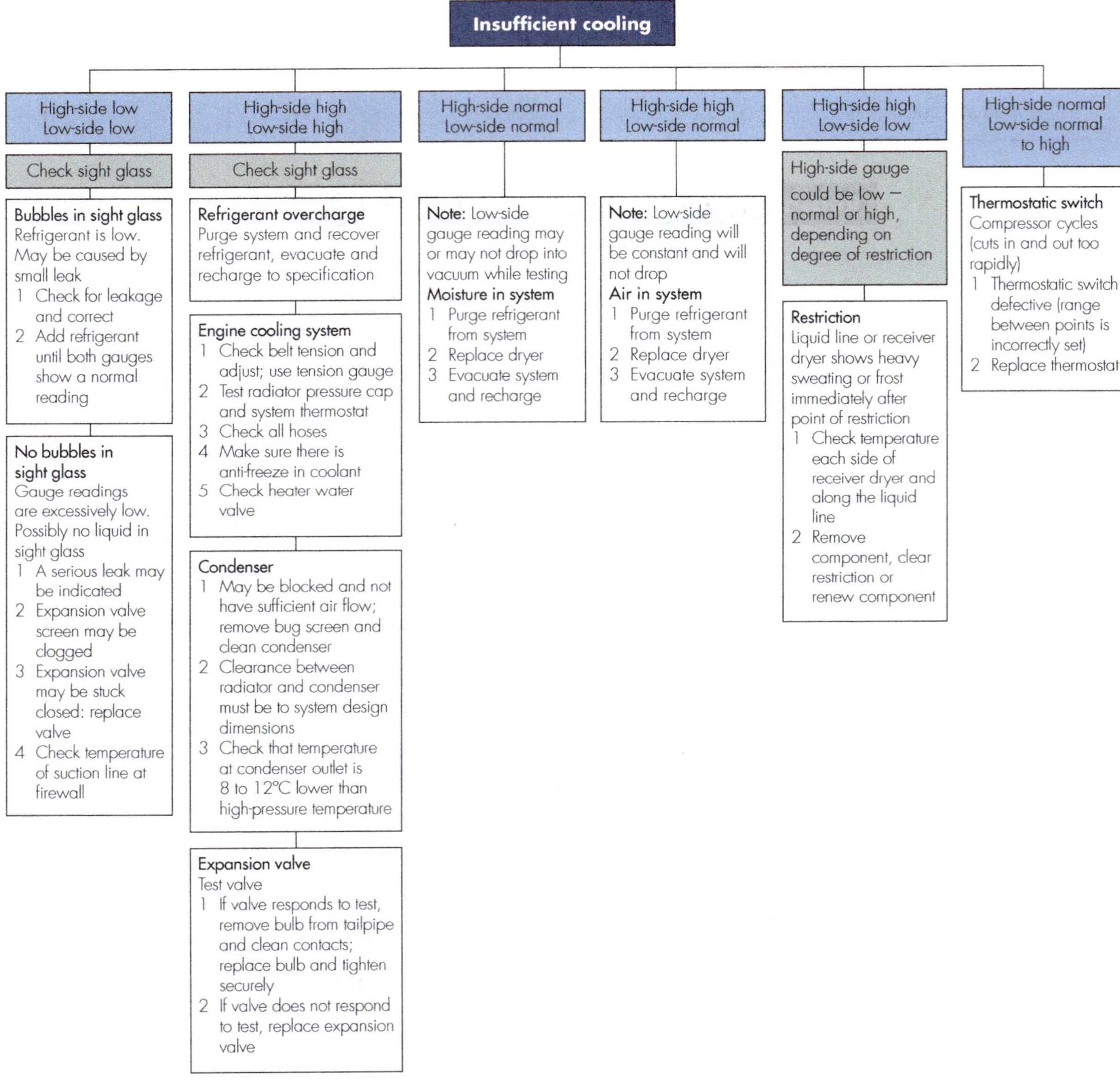

Source: Sanden International (Australia) Pty Ltd

System-diagnosis flow chart 3

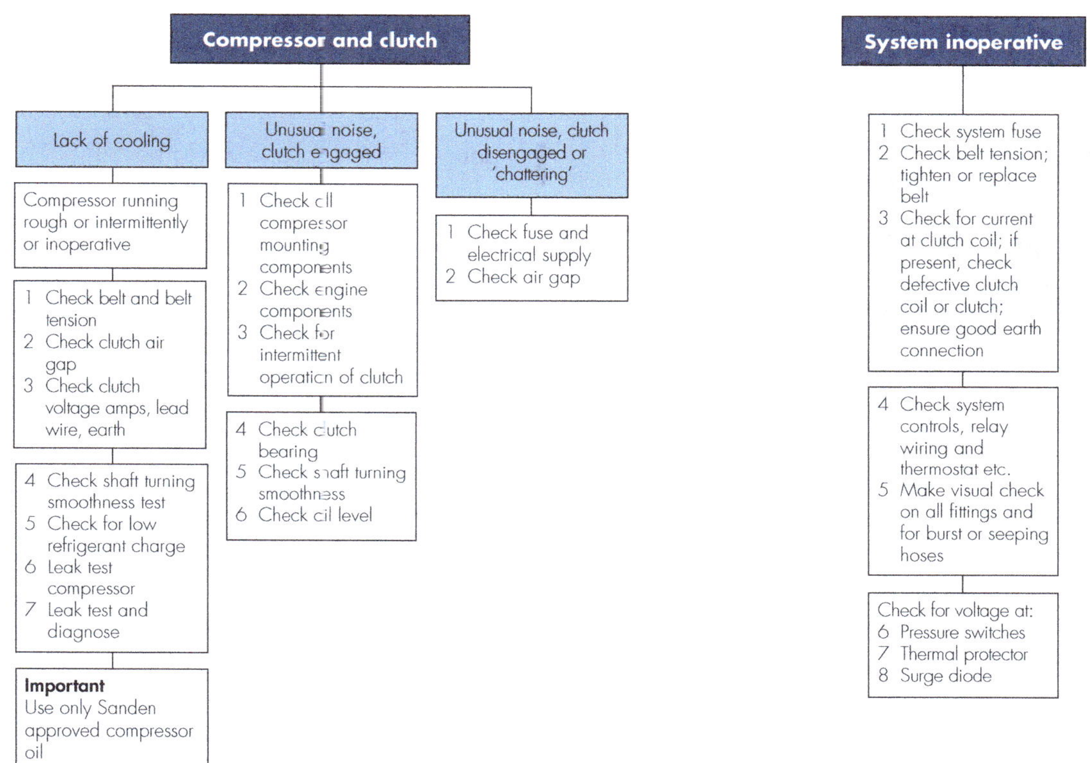

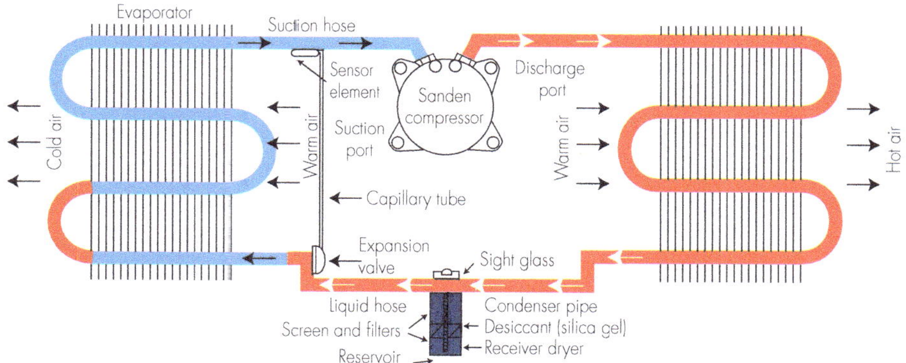

Source: Sanden International (Australia) Pty Ltd

2.2

System-diagnosis charts

Note: the ambient temperature in all system-diagnosis charts is 30°C. The system is designed for and uses R134a refrigerant

System-diagnosis chart A (normal refrigerant charge)

Chart A on page 153 shows the air-conditioning system working normally with all components showing average pressures.

The **compressor** will receive a cold but superheated refrigerant vapour of between 4 and 6°C, which cools the compressor in order to prevent overheating and seizure. The compressor then compresses this vapour, which in turn increases its pressure/temperature and its superheat value. The temperature of the refrigerant as it leaves the compressor is now above 75°C (depending on heat load).

The **high-pressure vapour line** transfers the high-pressure/temperature refrigerant to the condenser inlet. Its temperature is still above 75°C.

The **condenser** must first remove the extra heat that has been absorbed in the refrigerant. This is accomplished in the top third of the condenser, after which the condenser can start to condense the refrigerant or convert the refrigerant vapour into a liquid. This is done in the middle third of the condenser.

If a temperature probe were to be placed at this point, it would show that the temperature matched the pressure shown on the high-pressure gauge – in this case, it would be 58°C. **The high-pressure gauge therefore shows the condensing pressure/temperature of the system**, and not the superheated temperature of the refrigerant as it leaves the compressor and enters the condenser.

By the time that the refrigerant has reached the bottom of the condenser, it should mostly be a liquid and the refrigerant's temperature should have dropped by another 5 to 10°C under the pressure/temperature shown on the high-pressure gauge. This can be verified by placing a temperature probe on the condenser outlet and also using the gauge temperature scale.

The **receiver dryer** measures the temperature at both the receiver inlet and at the outlet. There should be no difference in temperature readings (as long as the receiver dryer is in good condition).

In the **high-pressure liquid line**, a refrigerant temperature drop of between 8 and 15°C below the actual gauge temperature (usually about 10 or 11°C) confirms that the system is fully charged.

As the refrigerant passes through the **TX valve**, its pressure (and therefore its temperature) drops as it enters the evaporator.

The temperature of the refrigerant is about –8°C as it enters the **evaporator**. The still-mostly-liquid refrigerant then starts to vaporise as it passes through the evaporator, and the temperature fluctuates as shown. The refrigerant should be 100 per cent vapour at the evaporator outlet. Complete vaporisation only occurs in the last section of the evaporator. Until then the refrigerant is considered to be a gaseous compound of liquid boiling into a vapour. At the evaporator outlet the temperature is between 0 and 2°C.

The **low-pressure vapour return line** returns the superheated vapour to the compressor, picking up heat from the engine as it goes to start its circulation all over again.

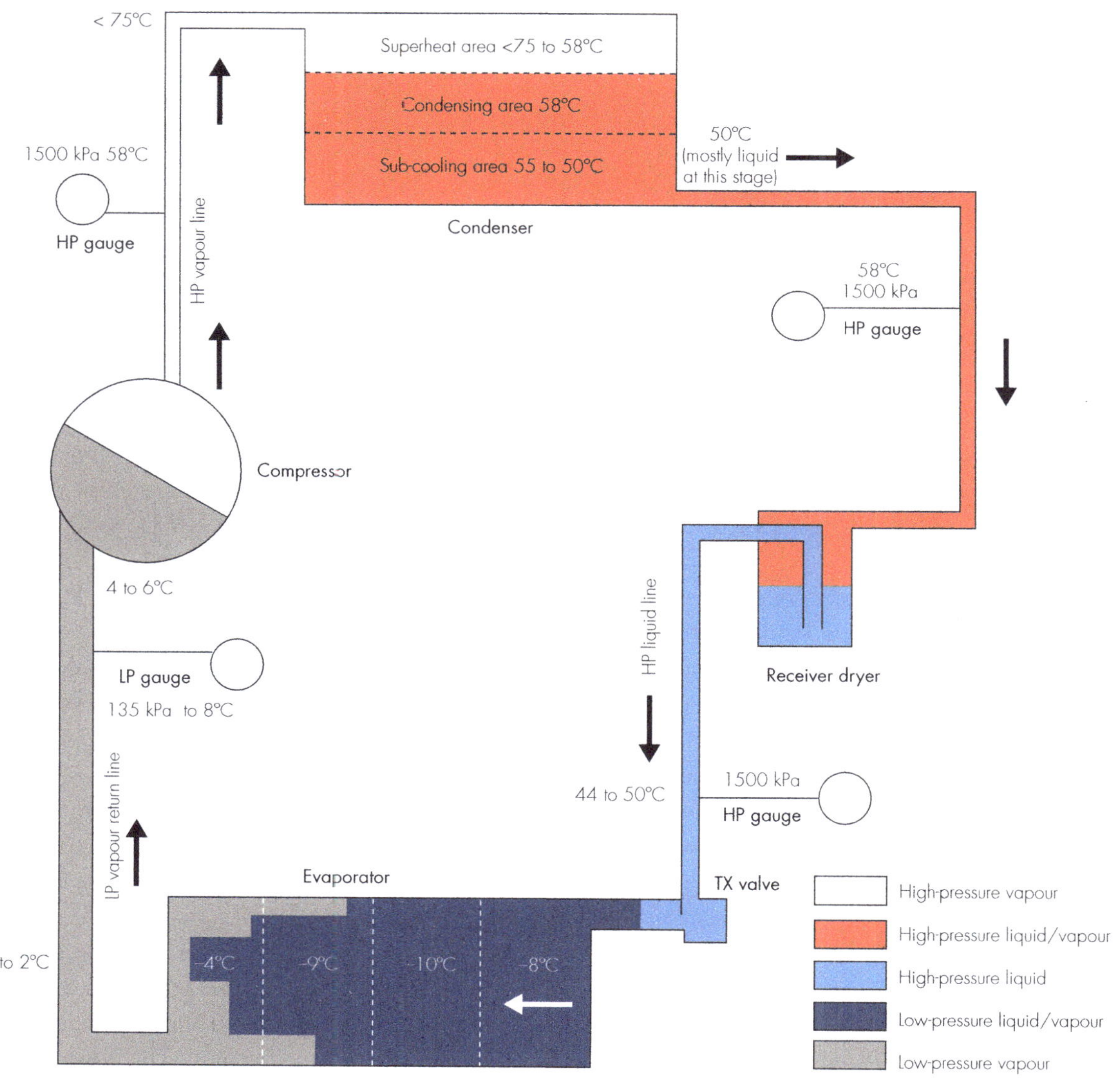

System-diagnosis chart A Normal operating temperatures and pressures

System-diagnosis chart B (low refrigerant charge)

Chart B on page 155 shows the air-conditioning system that has a low refrigerant charge with all components showing average pressures.

The **compressor** will receive a cool to warm superheated refrigerant vapour at a temperature of between 12 to 25°C, depending on how low the charge is. This vapour does not cool the compressor and can cause the superheat switch to activate in extreme conditions in order to prevent overheating and seizure. The compressor then compresses this vapour, which in turn increases its temperature and its superheat value rises dramatically. The temperature of the refrigerant as it leaves the compressor is now greater than 90°C.

The **high-pressure vapour line** transfers the high-pressure/temperature refrigerant to the condenser inlet. Its temperature is still above 90°C.

The **condenser** must first remove the extra heat that has been absorbed in the refrigerant, which is usually accomplished in the top third of the condenser. Notice the increased superheat that has developed in the system. Only when it is complete can the condenser start to condense the refrigerant or convert the refrigerant vapour into a liquid. However, this will take less effort as there is less refrigerant to condense.

If a temperature probe were to be placed at this point it would show that the temperature matches the pressure shown on the high-pressure gauge – 42°C in this case.

By the time that the refrigerant has reached the bottom of the condenser it should be mostly liquid, and its temperature will have dropped by only 0 to 4°C below the pressure/temperature shown on the high-pressure gauge. This can be verified by placing a temperature probe on the condenser outlet, and also by reading the gauge temperature scale.

The **receiver dryer** measures the temperature at both the receiver inlet and at the outlet. There should be no difference in temperature readings.

In the **high-pressure liquid line**, because this system problem is a low refrigerant charge, the temperature of the liquid line may show a very small drop. In some system designs it might even show an increase in temperature, depending on the amount of refrigerant that is missing from the system.

The refrigerant passes through the **TX valve**, and the pressure (and therefore the temperature) drops as it enters the evaporator. The TX valve is usually fully open in order to try to deliver enough refrigerant to the evaporator. It is therefore not modulating (low-pressure gauge fluctuating).

The temperature of the refrigerant as it enters the **evaporator** is approximately –8°C. The still-mostly-liquid refrigerant will quickly vaporise as it passes through the first section of the evaporator, but it then starts to take on superheat as it passes to the evaporator outlet.

This is called a *starved evaporator*. Vaporisation happens in the first section of the evaporator, resulting in a low cooling action. At the evaporator outlet the temperature is 8 to 10°C.

The **low-pressure vapour return line** returns the superheated vapour to the compressor, taking on extra superheat as it travels to once again begin its circulation.

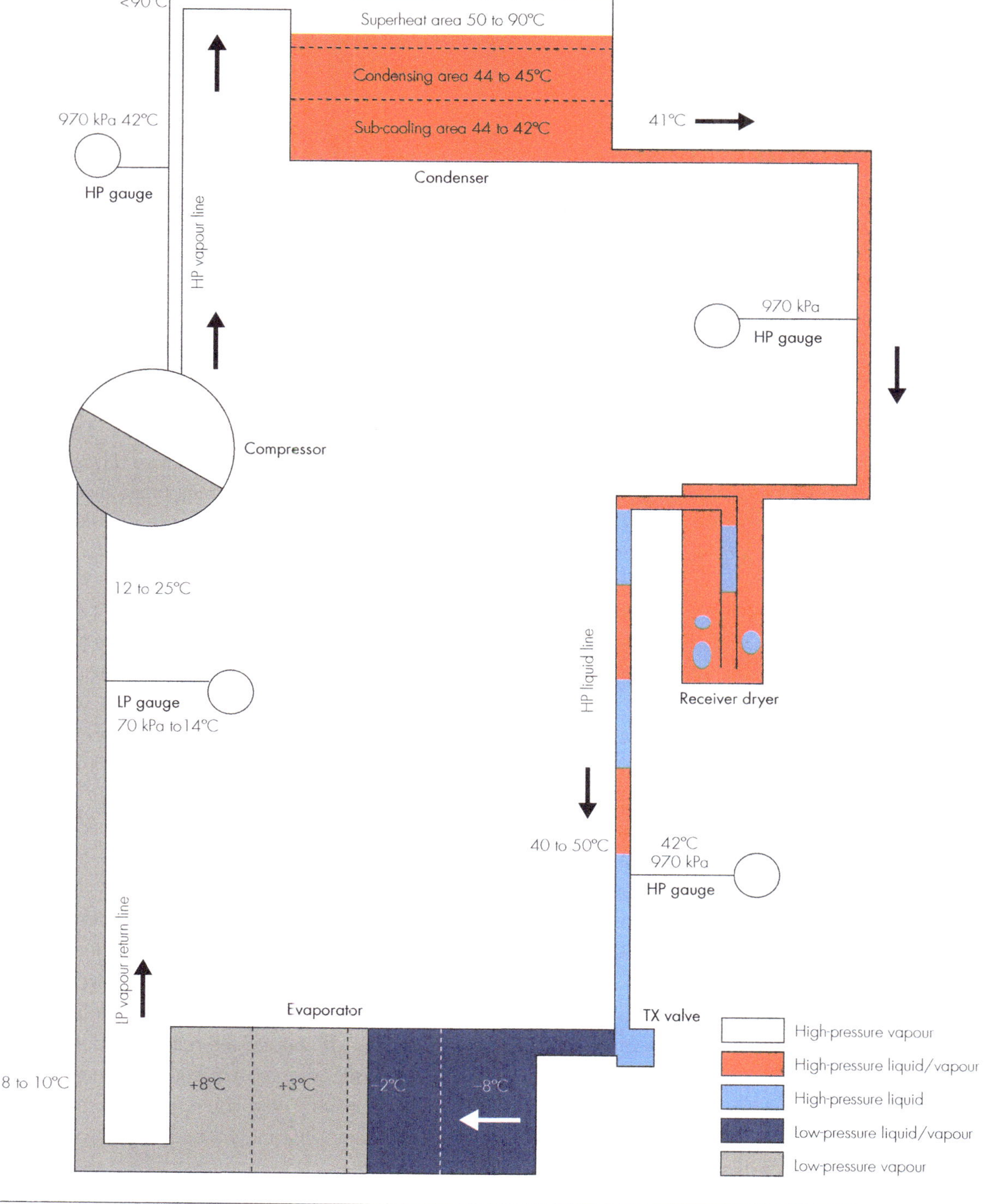

System-diagnosis chart B Low refrigerant charge

System-diagnosis chart C (system overcharged)

Chart C on page 157 shows the air-conditioning system with an **overcharge** of refrigerant and all components showing average pressures and temperatures.

The **compressor** will receive a very cold and slightly superheated refrigerant vapour of between −4 to −2°C and can have frosting of the return line almost to the compressor (possible traces of liquid). The compressor then compresses this vaporous mix, which in turn increases its temperature but only slightly increases its superheat value. The temperature of the refrigerant as it leaves the compressor is between 75 and 80°C (due to its cold state).

The **high-pressure vapour line** transfers the high-pressure/temperature refrigerant to the condenser inlet. Its temperature is still between 75 and 80°C.

The **condenser** must first remove the heat that has been absorbed in the refrigerant. The overcharge of refrigerant has allowed less superheated refrigerant in the condenser. This is accomplished in less than the top third of the condenser, after which the condenser can start to condense the refrigerant, or convert the refrigerant vapour into a liquid. This is done in the larger section of the condenser.

If a temperature probe were to be placed at this point it would show that the temperature matched the pressure shown on the high-pressure gauge, 70°C in this case.

By the time that the refrigerant has reached the bottom of the condenser, it should be a liquid, and its temperature should have dropped by between 5 and 8°C below the pressure/temperature shown on the high-pressure gauge. This can be verified by placing a temperature probe on the condenser outlet, and also by converting the gauge pressure to a temperature.

In the **receiver dryer** measure the temperature at both the receiver inlet and at the outlet. There should be no difference in the temperature readings.

In the **high-pressure liquid line** a temperature drop of more than 15°C shows that the system is overcharged.

The refrigerant passes through the **TX valve** and the pressure and temperature drops as it enters the evaporator. In this case it will be struggling to control the pressure and flow into the evaporator and therefore the suction pressure may be higher than normal.

The temperature of the refrigerant as it enters the **evaporator** is about −6°C. The still liquid refrigerant then passes through the evaporator and starts to vaporise slowly as shown. It is still vaporising as it reaches the evaporator outlet, and may not completely vaporise until it is in the return line or at the compressor. This is known as a *flooded evaporator*. The temperature is still −5°C at the evaporator outlet.

In the **low-pressure vapour return line** refrigerant is still vaporising in the return line as the flow through the evaporator exceeds the amount required for good evaporator operation. Therefore, the returning refrigerant has less superheated vapour and, under certain circumstances, some liquid, as it returns to the compressor where it will once again begin the circulation.

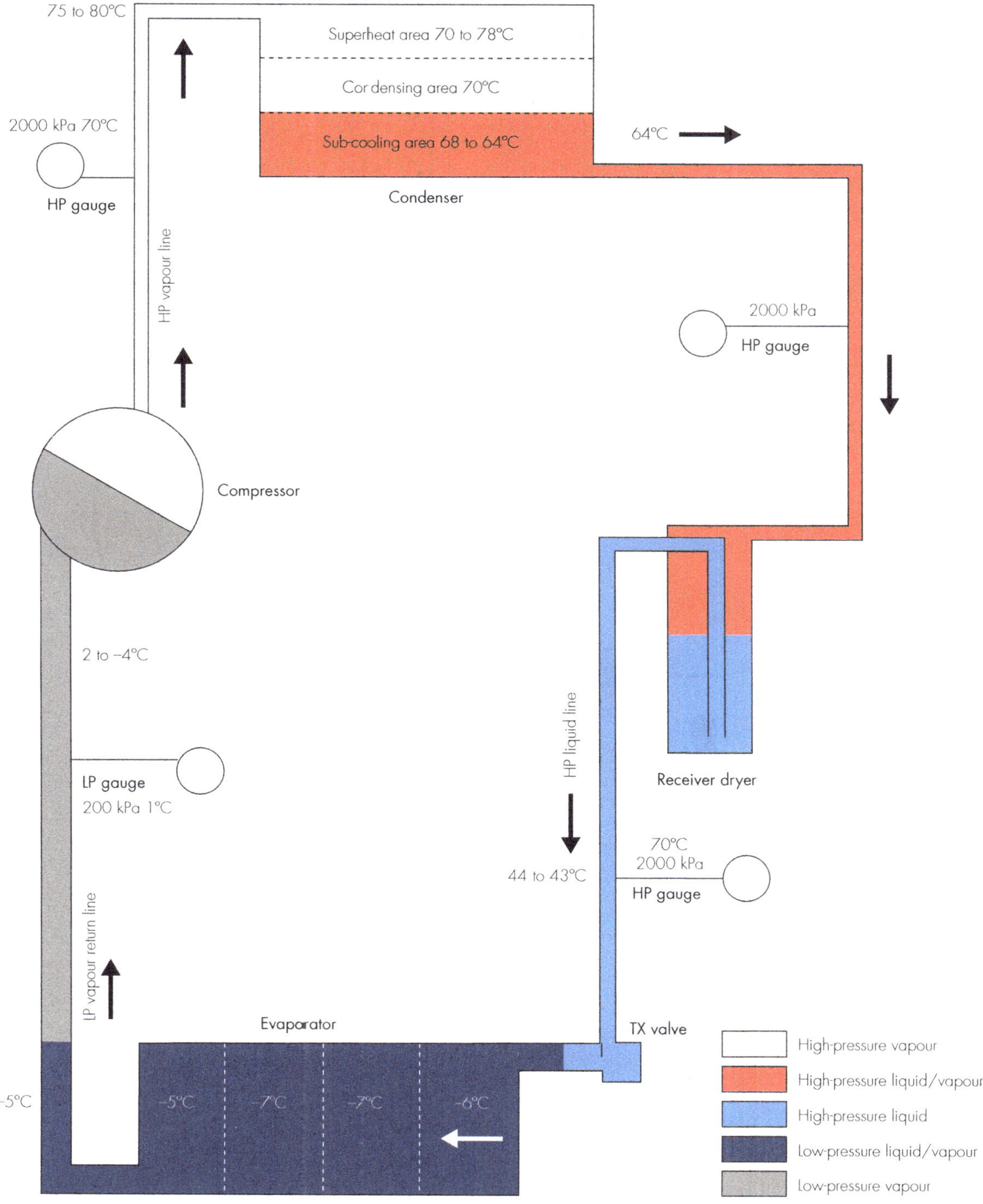

System-diagnosis chart C Overcharge of refrigerant

2.3

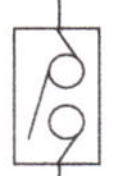

System-diagnosis problem sheets

Note: the ambient temperature for all diagnosis sheets is 30°C and all pressures relate to a cycling clutch system

System-diagnosis problem sheet 1

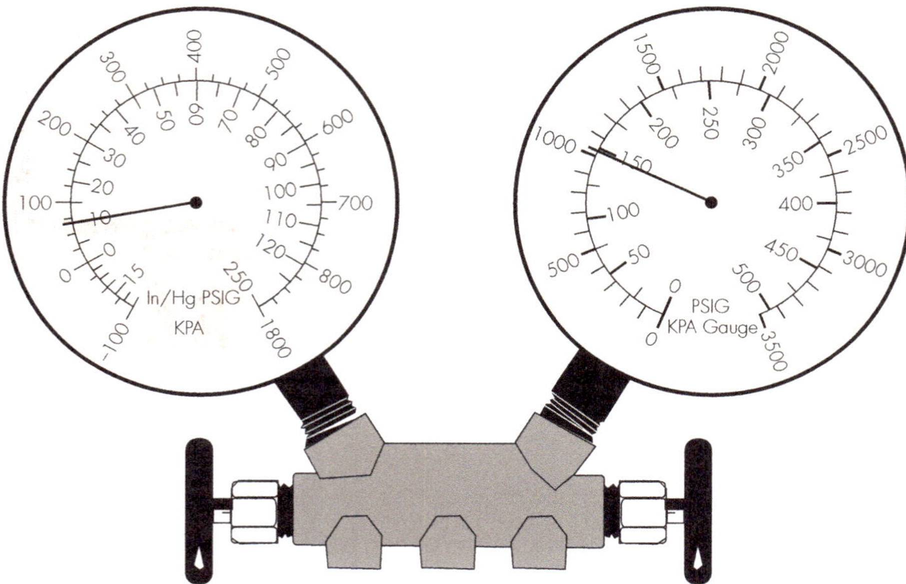

Low-gauge reading: Low
High-gauge reading: Slightly low
Complaint: Little or no cooling, especially on warmer days

System conditions:

1. Low-side gauge reading low – should read 100–180 kPa
2. High-side gauge reading slightly low – should read 1400–1600 kPa
3. Compressor cycling on time extended
4. Evaporator air cool but not cold
5. Low-pressure suction line temperature high (5 to 8°C)

Diagnosis: Low refrigerant charge – possible leak

Corrective steps:

1. Leak test the system
2. Purge and recover the refrigerant
3. Repair leak
4. Check the compressor oil level
5. Replace the receiver dryer
6. Evacuate the system and make sure the system holds a vacuum as per legislation
7. Recharge the system to specifications and retest for leaks
8. Performance test the system

System-diagnosis problem sheet 2

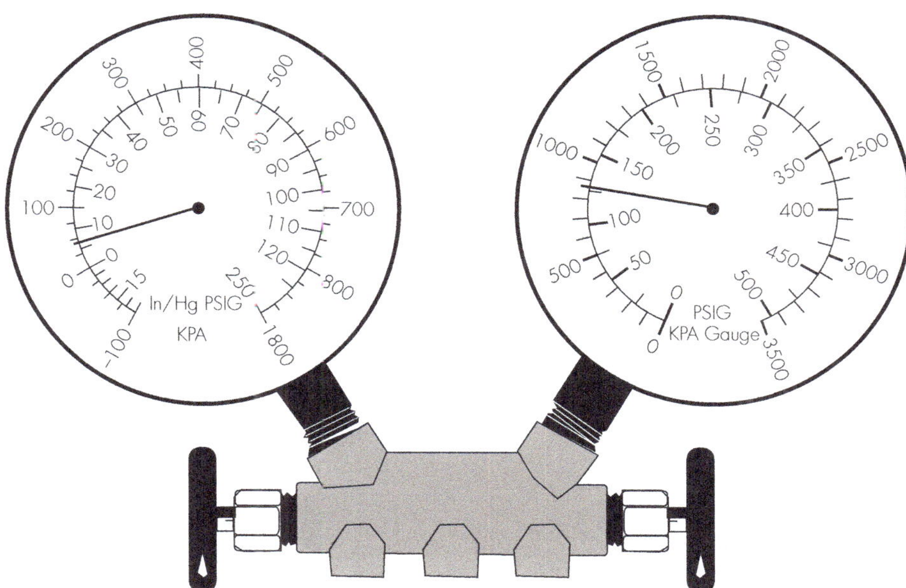

Low-gauge reading: Low
High-gauge reading: Low
Complaint: Little or no cooling, on cool and warmer days
System conditions:

1. Low-side gauge reading low – should read 100–180 kPa
2. High-side gauge reading low – should read 1400–1600 kPa
3. Low-pressure return line temperature warm (8 to 12°C)
4. Evaporator air cool but not cold
5. Compressor may stop (quick cycle times) if a low-pressure switch is fitted to the system, as low pressure drops below switch specification

Diagnosis: Extremely low refrigerant charge – serious leak
Corrective steps:

1. Leak test the system if not enough pressure in system to use an electronic detector (refer to leak testing chapter for other alternatives)
2. Purge and recover the refrigerant
3. Repair the leak
4. Check the compressor oil level
5. Replace the receiver dryer
6. Evacuate the system and make sure the system holds a vacuum
7. Recharge the system to specifications and retest for leaks
8. Performance test the system

System-diagnosis problem sheet 3

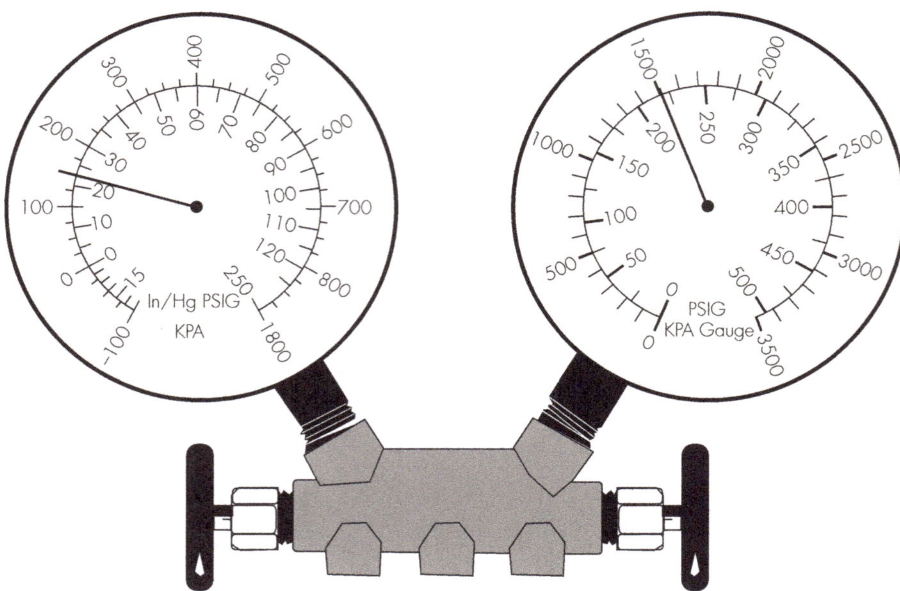

Low-gauge reading: High side of normal to slightly high

High-gauge reading: High side of normal to slightly high

Complaint: Little or no cooling

System conditions:

1 Low-side gauge reading normal to slightly high
2 High-side gauge reading normal to slightly high
3 Temperature at condenser inlet contains extra superheat
4 Evaporator air is cool but not cold; low-pressure return line temperature high (should be 0 to 2°C)
5 Low-side gauge does not change during cycle on; pressure should fluctuate during cycle on

Diagnosis: Air present in system – system also not fully charged

Corrective steps:

1 Leak test the system, giving compressor seal area special attention
2 Purge and recover the refrigerant
3 Repair system
4 Check the compressor oil level
5 Replace the receiver dryer
6 Evacuate the system and make sure the system holds vacuum
7 Recharge the system to specifications and retest for leaks
8 Performance test the system

System-diagnosis problem sheet 4

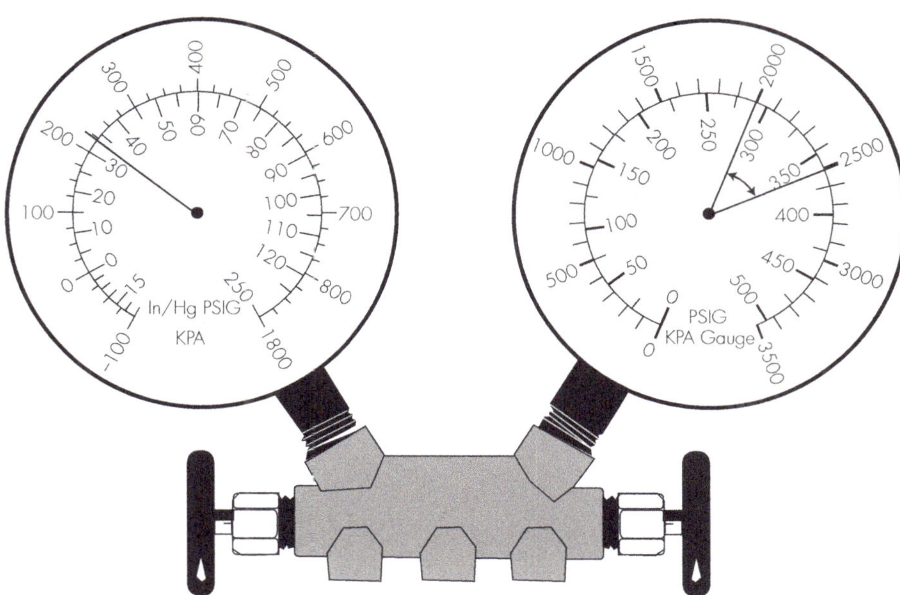

Low-gauge reading: High
High-gauge reading: High
Complaint: Little or no cooling
System conditions:

1 Low-side gauge reading high
2 High-side gauge reading high
3 Temperature at condenser inlet higher than normal, very hot to touch superheated (20 to 50 per cent)
4 Evaporator air cool but not cold; low-pressure return line temperature high (should be 0 to 2°C)
5 Compressor cycle on time extended

Diagnosis: Large amounts of air present in system, system is fully charged – system possibly not evacuated correctly

Corrective steps:

1 Leak test the system
2 Purge and recover the refrigerant
3 Check the compressor oil level
4 Replace the receiver dryer
5 Evacuate the system
6 Recharge the system to specifications and retest for leaks
7 Performance test the system

System-diagnosis problem sheet 5

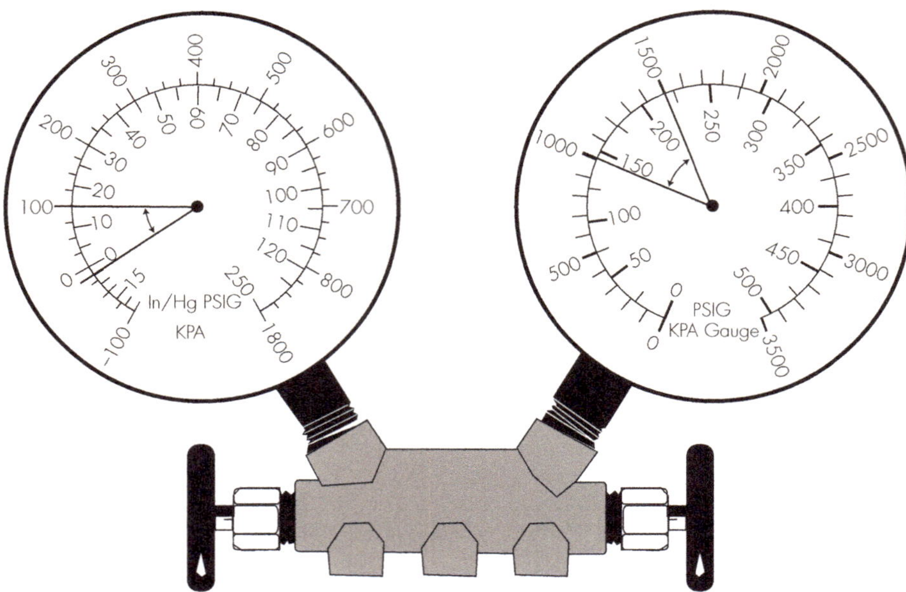

Low-gauge reading: Normal, then low, then extremely low or into a vacuum

High-gauge reading: Normal, then low as low-pressure drops

Complaint: Little or no cooling after 5–15 minutes, especially around midday on warmer days

System conditions:

1 Low-side gauge reading normal then low
2 High-side gauge reading normal then low
3 Low pressure return line temperature fluctuates 0–2°C then up to as much as 25°C
4 Evaporator air cold, but becomes warm as the low-side gauge moves very low. As low-side drops, the high-side also drop.
5 As the low-pressure drops, the low-pressure switch activates, affecting compressor cycle times

Diagnosis: Excessive moisture is present in the system, the system is fully charged

Corrective steps:

1 Leak test the system
2 Purge and recover the refrigerant
3 Check the compressor oil level
4 Replace the receiver dryer
5 Evacuate the system for extended time
6 Recharge the system to specifications and retest for leaks
7 Performance test the system

System-diagnosis problem sheet 6

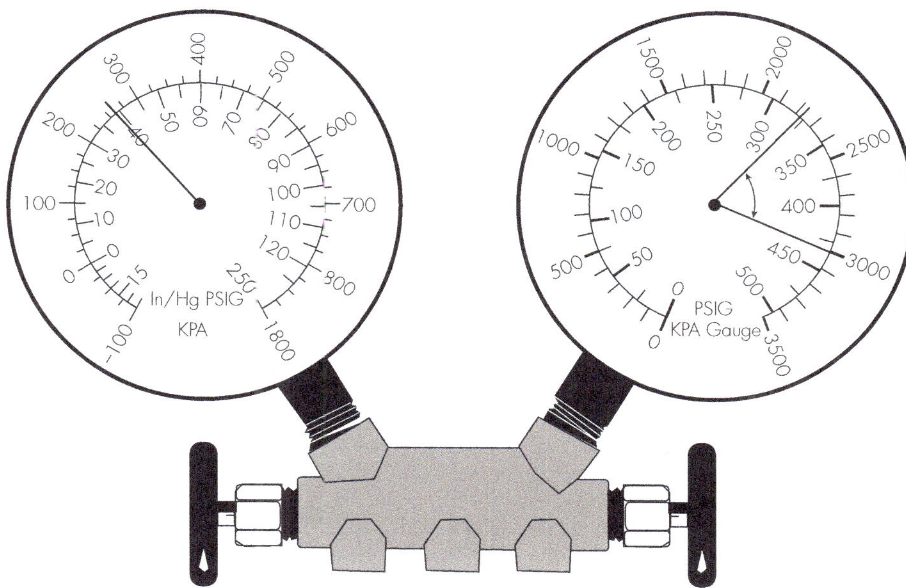

Low-gauge reading: High
High-gauge reading: High
Complaint: Decreased cooling efficiency; engine may also overheat on hotter days

System conditions:

1. Low-side gauge reading high
2. High-side gauge reading high – high-pressure switch may activate
3. Liquid line temperature higher than specified
4. Evaporator air warmer than normal
5. Superheat very high at condenser inlet
6. Condenser outlet temperature high

Diagnosis: Lack of cooling by improper condenser operation (air flow restricted) – fans not working or the system could be overcharged

Corrective steps:

1. Check operation of engine fan and auxiliary fans
2. Inspect condenser for external obstructions
3. Inspect high-pressure lines from the compressor for faults and or damage
4. Check system for overcharge by:
   - recovering and measuring the amount of refrigerant
   - replace the receiver dryer
   - checking the compressor oil level
   - evacuating the system for extended time
   - recharging the system to specifications and test for leaks
   - performance testing the system
5. If fault still exists, remove condenser and check for internal obstructions and or dirt

System-diagnosis problem sheet 7

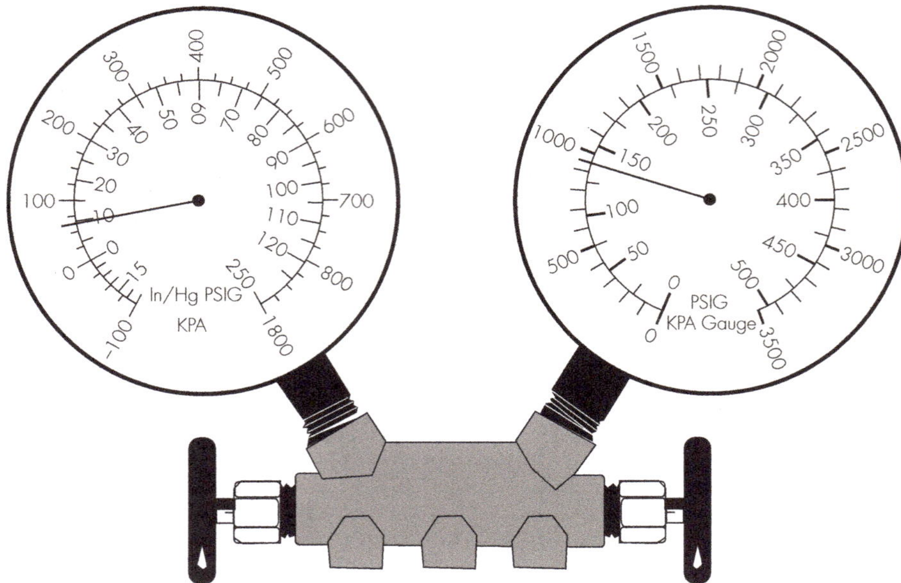

Low-gauge reading: Normal then low
High-gauge reading: Normal then low
Complaint: Decreased cooling efficiency
System conditions:

1. Low-side gauge reading normal then drops
2. High-side gauge reading – as the low pressure drops the discharge pressure also starts to drop
3. High-pressure liquid line temperature is lower than specified, and may show sweating (depending on the amount of restriction)
4. Evaporator air warmer than normal – low-pressure return line temperature rises

Diagnosis: Restriction of the liquid line or receiver dryer
Corrective steps:

1. Purge and recover the refrigerant
2. Check the compressor oil level
3. Replace any line that shows frosting in one point; this is a restriction (not on FOT systems as this is normal)
4. Replace the receiver dryer
5. Evacuate the system
6. Recharge the system to specifications and test for leaks
7. Performance test the system

System-diagnosis problem sheet 8

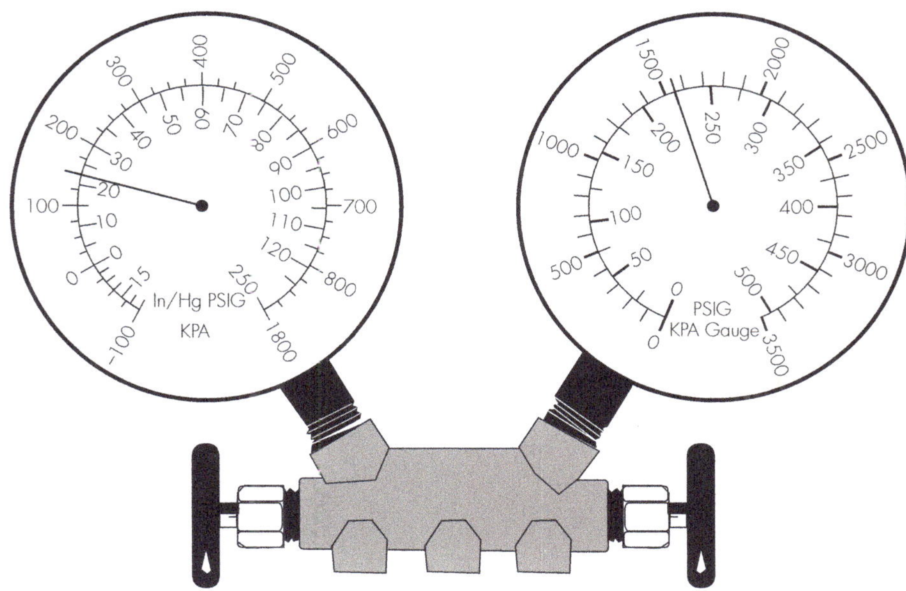

Low-gauge reading: Normal, but can be slightly lower than normal

High-gauge reading: Normal

Complaint: Compressor stays on nearly all the time, frosting on the suction return pipe nearly all the time or vent temperature very cold

System conditions:

1 Low-side gauge reading normal to low
2 High-side gauge reading normal
3 Compressor cycles on for long periods of time vent temperature very cold
4 Evaporator may freeze if the compressor cycles on for too long, restricting the air flow

Diagnosis: Thermostatic switch out of adjustment set too low. Sensor pipe broken or dislodged inside evaporator

Corrective steps:

1 Remove the covers to gain access to the thermostatic assembly and (1) check the positioning of the sensor pipe; (2) gain access to the adjusting switch if it has one; and (3) adjust to cut off compressor at approximately 4 to 6°C
2 Complete the performance testing of the system
3 Re-install all covers

System-diagnosis problem sheet 9

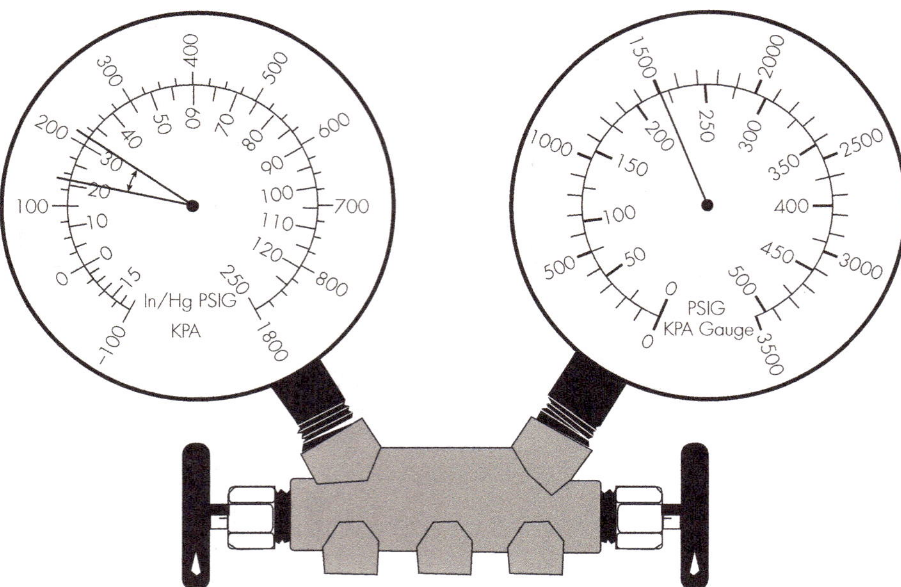

Low-gauge reading: Normal to slightly high
High-gauge reading: Normal
Complaint: Compressor cycles too rapidly; insufficient cabin cooling on hot days
System conditions:

1 Low-side gauge reading normal to slightly high
2 High-side gauge reading normal
3 Compressor cycles on and off quickly
4 Low-side gauge does not reach its correct pressure before the compressor cycles off

Diagnosis: Thermostatic switch defective (temperature difference sensor faulty). Pressure switches faulty

Corrective steps:

1 Some thermostatic switches are fitted with a sensitivity adjustment. If this is available, check the adjustment from the manufacturer's specifications.
2 If no adjustment is available, replace the switch and then performance test the system.
3 Bypass the high and or low-pressure switch and check operation of system.

NOTE

+ For this last stage in the corrective steps, keep close attention to the pressure gauges while testing the system and remember to remove the bypass when testing is complete.

I System-diagnosis problem sheet 10

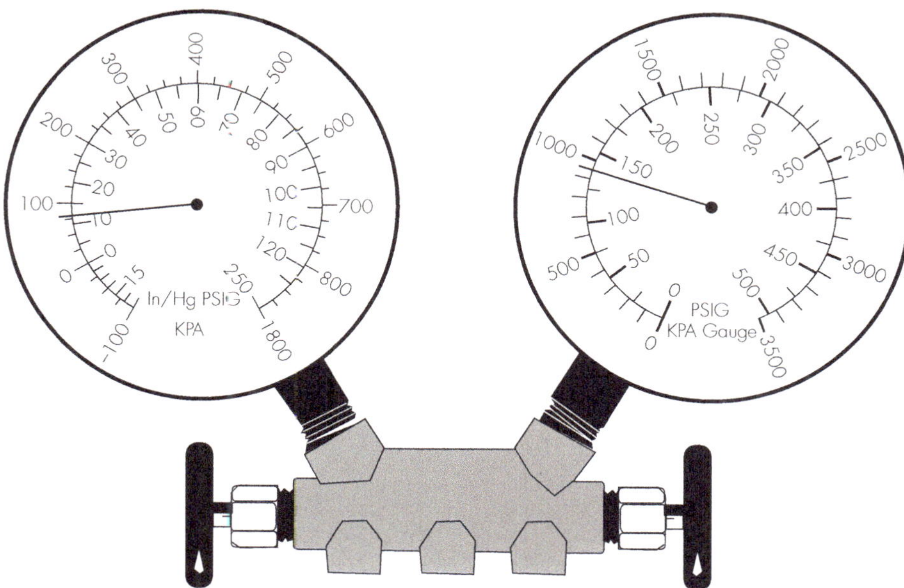

Low-gauge reading: Normal then low
High-gauge reading: Normal then low
Complaint: Insufficient cooling
System conditions:

1 Low-side gauge reading normal then low – return line temperature high
2 High-side gauge reading drops to low as suction pressure drops
3 Expansion valve is sweating or is frosted (evaporator is *starved*)
4 Evaporator outlet air is cool but not cold
5 Low-pressure switch may activate

Diagnosis: Expansion valve is stuck closed, or clogged
Corrective steps:

1 Disconnect the compressor clutch and wait until the frosting has gone. Reconnect the clutch and if the valve behaves itself for a few minutes but reverts to the problem, moisture is in the system. If no change in operation replace the expansion valve
2 Purge and recover the refrigerant
3 Replace the expansion valve
4 Check compressor oil and replace the receiver dryer
5 Evacuate the system for an extended time
6 Recharge the system to specification and teak test
7 Performance test the system

System-diagnosis problem sheet 11

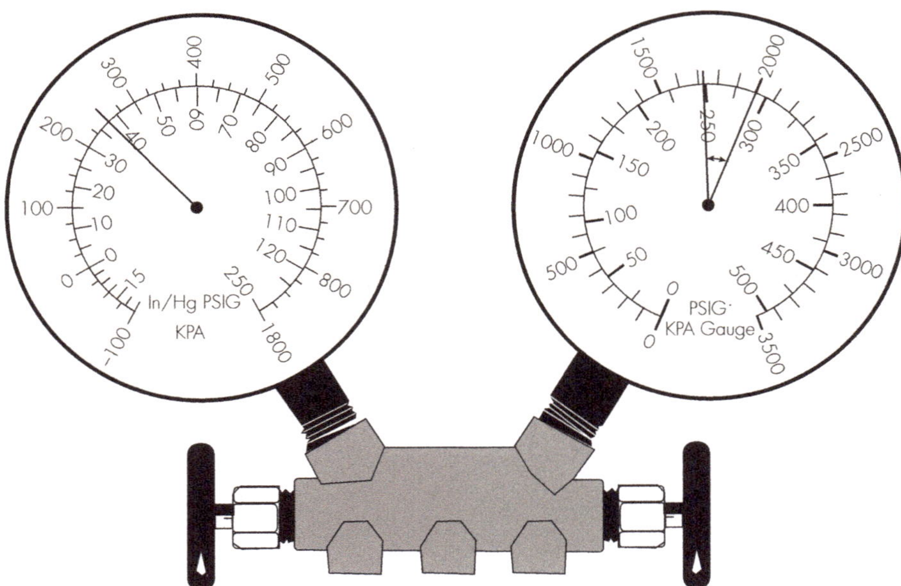

Low-gauge reading: High
High-gauge reading: High
Complaint: Insufficient or no cooling
System conditions:

1 Low-side gauge reading high
2 High-side gauge reading high – superheat normal
3 Evaporator and suction return line is sweating or is frosted (evaporator is flooded)
4 Evaporator outlet pipe shows considerable moisture; temperature below 0°C
5 Air discharge is cool only 8–12°C

Diagnosis: Expansion valve is stuck open
Corrective steps:

1 Inspect the positioning of the expansion valve sensing coil if out of position, clean and replace in position. If fault still exists, expansion valve is jammed open and needs to be replaced
2 Replace the valve as follows:
3 Purge and recover the refrigerant; replace the expansion valve
4 Check the compressor oil and replace the receiver dryer
5 Evacuate the system for an extended time
6 Recharge the system to specification and leak test
7 Performance test the system

System-diagnosis problem sheet 12

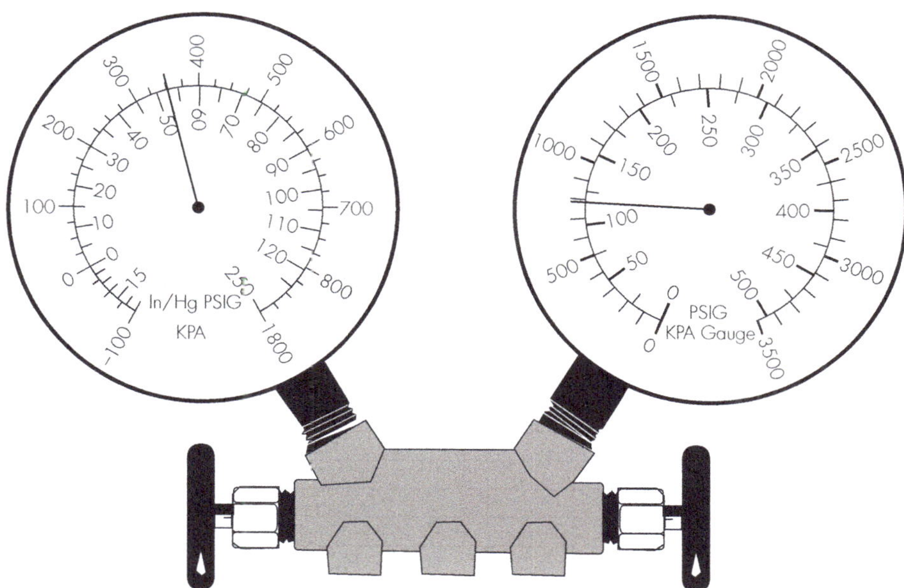

Low-gauge reading: High
High-gauge reading: Low
Complaint: Little or no cooling
System conditions:

1. Low-side gauge reading high
2. High-side gauge reading low
3. Compressor cycles in continuously
4. Compressor can vibrate or be noisy in operation (piston, rings and/or valves)
5. Pressure gauge readings can vibrate (valves)
6. Evaporator outlet air cool only

Diagnosis: Compressor malfunction (lack of suction and/or pumping action)

Corrective steps:

Internal fault indicated; repair compressor if parts for that compressor are available or replace the compressor

1. Purge and recover the refrigerant
2. Remove and overhaul if parts are available or replace the compressor
3. Replace the receiver dryer
4. Evacuate and recharge the system, then leak test
5. Performance test the system

2.4

Pressure/temperature charts

Note: all pressures are shown in kPa

System operating pressure/temperature chart

Ambient temperature (°C)	R134a head pressure	R1234yf head pressure	Centre vent air temperature (°C)
15	700–850	675–800	0–4
16	700–850	675–825	0–4
17	750–900	725–900	1–4
18	800–1000	775–1000	1–5
19	900–1100	850–1050	2–6
20	1050–1200	950–1100	2–7
21	1050–1225	950–1125	3–8
22	1075–1250	975–1150	3–8
23	1100–1300	1000–1175	4–8
24	1100–1350	1000–1200	4–8
25	1130–1400	1030–1240	4–9
26	1150–1450	1050–1280	5–9
27	1200–1500	1075–1300	5–10
28	1225–1550	1125–1350	5–10
29	1275–1600	1175–1420	6–11
30	1350–1650	1200–1460	6–12
31	1400–1650	1250–1480	6–12
32	1450–1750	1275–1480	7–13
33	1500–1800	1300–1500	7–14
34	1550–1850	1375–1550	8–14
35	1600–1950	1400–1550	9–15
36	1650–2000	1425–1600	10–15
37	1700–2100	1475–1650	10–15
38	1750–2150	1500–1650	11–15
39	1800–2200	1550–1700	11–16
40	1800–2300	1600–1750	12–16
41	1900–2400	1600–1775	12–16
42	2000–2450	1650–1800	12–17
43	2100–2550	1700–1900	13–17
44	2200–2700	1800–2000	13–17
45	2300–2800	1900–2150	13–18

*Use this chart only as a guide (actual pressures may vary according to the system).

*Make sure that there is no high-pressure gauge creep (i.e., the system is handling the heat load).

*Note that R1234yf head pressure is generally 100 kPa lower than R134a.

Refrigerant pressure/temperature chart

R134a (kPa)	R1234yf (kPa)	°C
10	27.4	-24
20.3	38.7	-22
31.4	49.5	-20
43.2	61.6	-18
55.9	74.5	-16
69.4	88.7	-14
83.8	102.5	-12
99.1	117.7	-10
115.4	133.3	-8
132.7	150.7	-6
151.1	168.5	-4
170.6	187.4	-2
191.2	207.1	0
213	227.9	2
236	249.3	4
260.2	272.7	6
285.8	296.7	8
312.8	321.3	10
341.2	348.1	12
371	375.5	14
402.4	404.4	16
435.2	434.4	18
469.7	465.7	20
505.9	498.4	22
543.8	532.5	24
583.4	567.9	26
624.8	604.9	28
668.2	643.3	30
713.4	683.2	32
760.6	724.7	34
809.8	767.8	36
861.2	812.5	38
914.6	859	40
970.3	907.1	42
1028.3	957	44
1088.5	1008.7	46
1151.2	1062.2	48
1216.3	1117.6	50
1283.9	1174.9	52
1354.1	1234.1	54
1426.9	1295.4	56

R134a (kPa)	R1234yf (kPa)	°C
1502.4	1358.7	58
1580.7	1424.160	60
1661.9	1491.662	62
1789.1	1597	65
2016.6	1783.9	70
2263.8	1985.575	75
2532.6	2202.6	80
2824.3	2435.9	85
3140.7	2686.4	90
3484.5	2954.7	95

The shaded area indicates the ideal low-pressure range for a correctly working system.

2.5

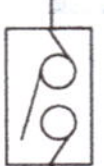

Temperature–pressure exercises

Note: You will need to use the pressure/temperature charts on the previous pages to complete these exercises. Consider all conditions as normal

Temperature–pressure relationship exercise 1

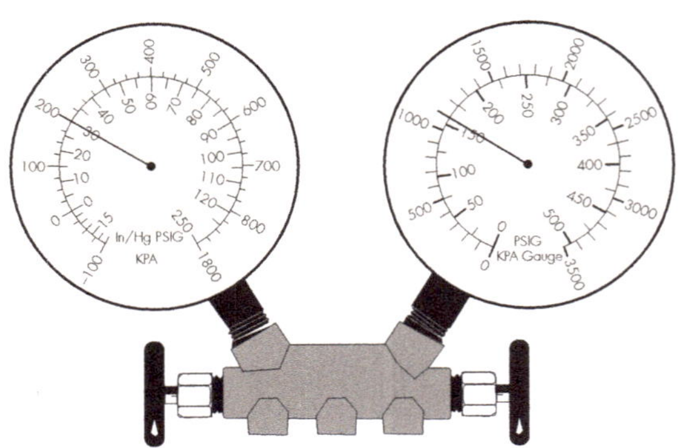

1 What approximate pressure is indicated on the low-side gauge? ______________ **kPa**
2 This pressure corresponds to an evaporator temperature of ______________ **°C**
3 What approximate pressure is indicated on the high-side gauge? ______________ **kPa**
4 This pressure corresponds to an ambient temperature of ______________ **°C**
5 What will be the temperature of the air recorded at the centre vent outlet?
Minimum ______________ **°C**
Maximum ______________ **°C**

Temperature–pressure relationship exercise 2

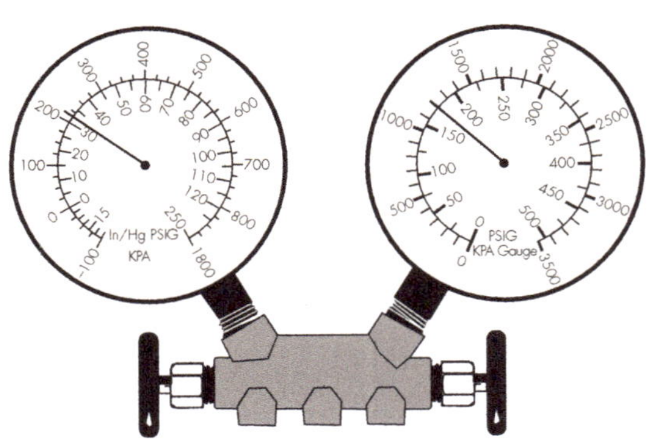

1 What approximate pressure is indicated on the low-side gauge? ______________ **kPa**
2 This pressure corresponds to an evaporator temperature of ______________ **°C**
3 What approximate pressure is indicated on the high-side gauge? ______________ **kPa**
4 This pressure corresponds to an ambient temperature of ______________ **°C**
5 What will be the temperature of the air recorded at the centre vent outlet?
Minimum ______________ **°C**
Maximum ______________ **°C**

Temperature–pressure relationship exercise 3

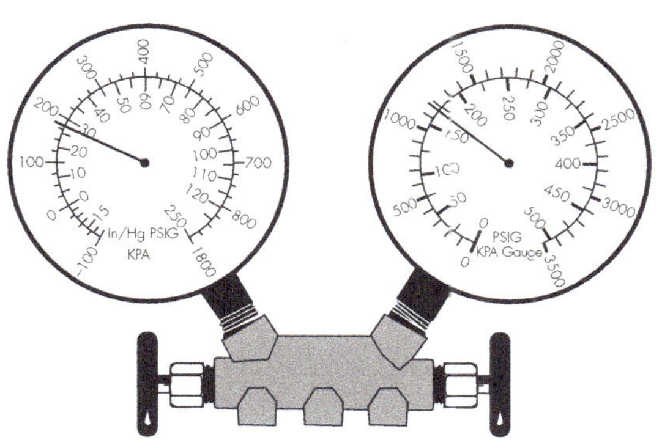

1 What approximate pressure is indicated on the low-side gauge? ______ kPa
2 This pressure corresponds to an evaporator temperature of ______ °C
3 What approximate pressure is indicated on the high-side gauge? ______ kPa
4 This pressure corresponds to an ambient temperature of ______ °C
5 What will be the temperature of the air recorded at the centre vent outlet?
Minimum ______ °C
Maximum ______ °C

Temperature–pressure relationship exercise 4

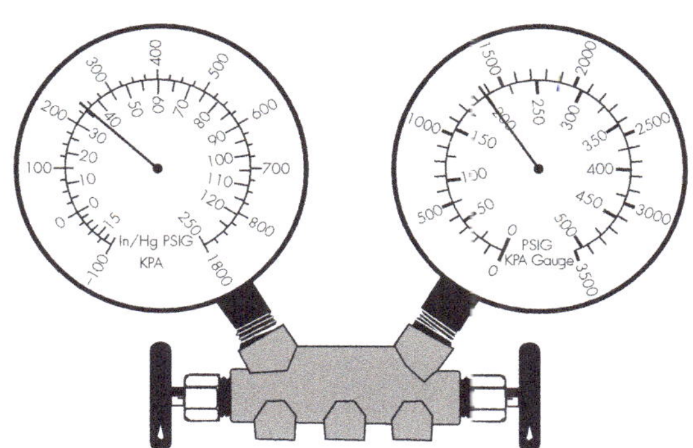

1 What approximate pressure is indicated on the low-side gauge? ______ kPa
2 This pressure corresponds to an evaporator temperature of ______ °C
3 What approximate pressure is indicated on the high-side gauge? ______ kPa
4 This pressure corresponds to an ambient temperature of ______ °C
5 What will be the temperature of the air recorded at the centre vent outlet?
Minimum ______ °C
Maximum ______ °C

Temperature–pressure relationship exercise 5

1 With an ambient temperature of 35°C, what is the minimum head pressure?

_______________ kPa

2 With an evaporator temperature of −3°C, what is the suction pressure?

_______________ kPa

Temperature–pressure relationship exercise 6

1 A minimum head pressure is 1275 kPa. What is the ambient temperature?

_______________ °C

2 If the evaporator temperature is 0°C, what is the low-side gauge reading?

a _______________ kPa

Temperature–pressure relationship exercise 7

1 The ambient temperature is 38°C. What is the normal high-side pressure between?

_______________ kPa

2 The low-side pressure 115 kPa. What is the temperature of the refrigerant in the evaporator? _______________ °C

3 The centre outlet air temperature in this problem will be

Minimum _______________ °C

Maximum _______________ °C

Temperature–pressure relationship exercise 8

1 The minimum high-side gauge reading is 1345 kPa. What will be the ambient temperature in this problem?

_______________ °C

2 The low-side gauge reads 150 kPa. What is the evaporator temperature in this problem?

_______________ °C

3 The centre outlet air temperature in this problem will be

Minimum _______________ °C

Maximum _______________ °C

2.6

System-diagnosis exercises

Note: You will need to use the pressure/temperature charts on the previous pages to complete these exercises. Consider all conditions as normal

System-diagnosis exercise 1

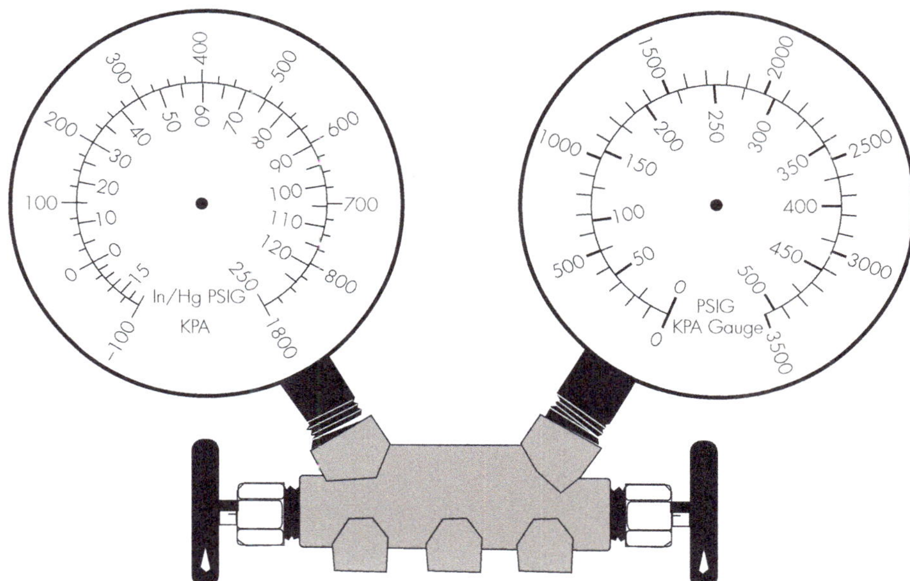

The compressor – cycling clutch TX valve or FOT system

Conditions

- Ambient temperature: 32°C
- Suction (low) pressure: 345 kPa
- Head (high) pressure: 830 kPa

Diagnosis

1 On the above manifold gauge diagram draw in the needles to indicate the pressures.

2 For this ambient temperature the normal head pressure is ________________ **kPa**

3 Refrigerant evaporator temperature for 345kPa is ________________ **°C**

4 The suction pressure for this system is **low/normal/high** ________________

5 The head pressure for this system is **low/normal/high** ________________

6 These conditions result in the evaporator cooling being **good/poor** ________________

7 The air temperature at the centre vent outlet will be ________________ **-4–0°C/4–6°C/8–10°C/12–14°C**

8 The compressor is noisy in operation; this indicates (from diagnostic sheets) ________________

____

I System-diagnosis exercise 2

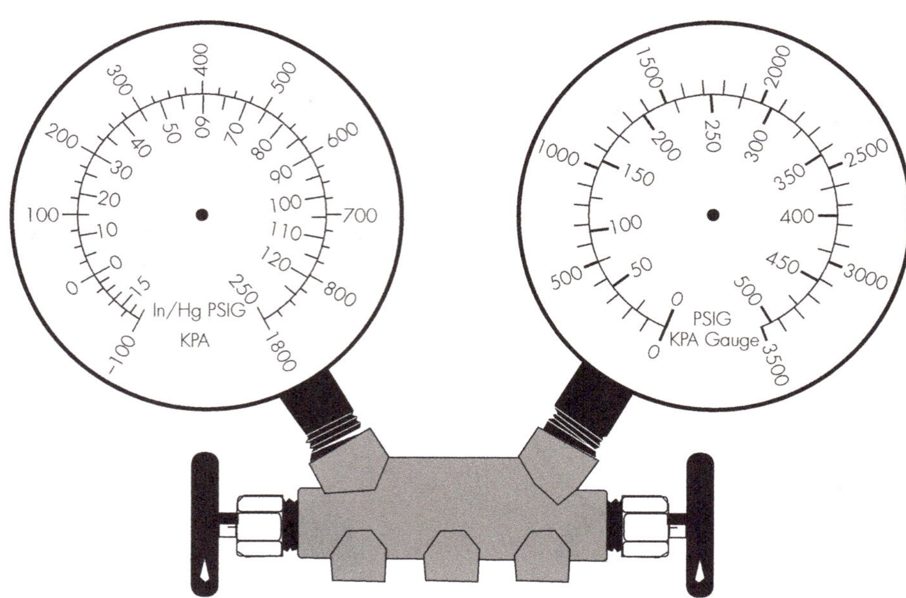

The compressor – cycling clutch TX valve, or FOT system

Conditions

- Ambient temperature: 35°C
- Suction (low) pressure: 395 kPa
- Head (high) pressure: 2500 kPa

Diagnosis

1 On the above manifold gauge diagram draw in the needles to indicate the pressures.

2 For this ambient temperature the normal head pressure is ______________ **kPa**

3 Refrigerant evaporator temperature for 395kPa is ______________ **°C**

4 The suction pressure for this system is **low/normal/high** ______________

5 The head pressure for this system is **low/normal/high** ______________

6 These conditions result in the evaporator cooling being **good/very poor** ________

7 Give two external faults that could cause this head pressure:

a ______________________________

b ______________________________

8 Give one internal fault that could cause this head pressure:

a ______________________________

System-diagnosis exercise 3

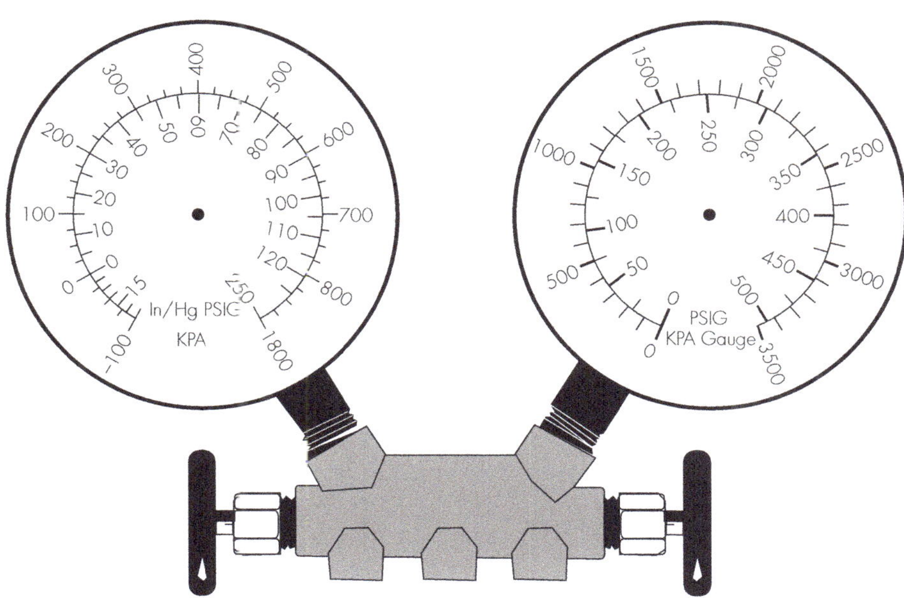

The receiver dryer – cycling clutch TX valve system

Conditions

- Ambient temperature: 38°C
- Suction (low) pressure: 40 kPa
- Head (high) pressure: 1800 kPa

Diagnosis

1 On the above manifold gauge diagram draw in the needles to indicate the pressures.

2 For this ambient temperature the normal head pressure is ______________ **kPa**

3 Normal suction pressure for this ambient temperature should be ______________ **kPa**

4 Refrigerant evaporator temperature for 40 kPa is ______________ **°C**

5 The head pressure for this system is **low/normal/high** ______________

6 These conditions result in the evaporator cooling being **good/poor** ______________

Give reasons for your answer ______________

____

____

7 The air temperature at the centre vent outlet will be ______________

–4 to 0/4 to 6/8 to 10/12 to14°C

8 If a temperature difference across the receiver dryer is noticed:

a What is the fault? ______________

b What action should be taken? ______________

I System-diagnosis exercise 4

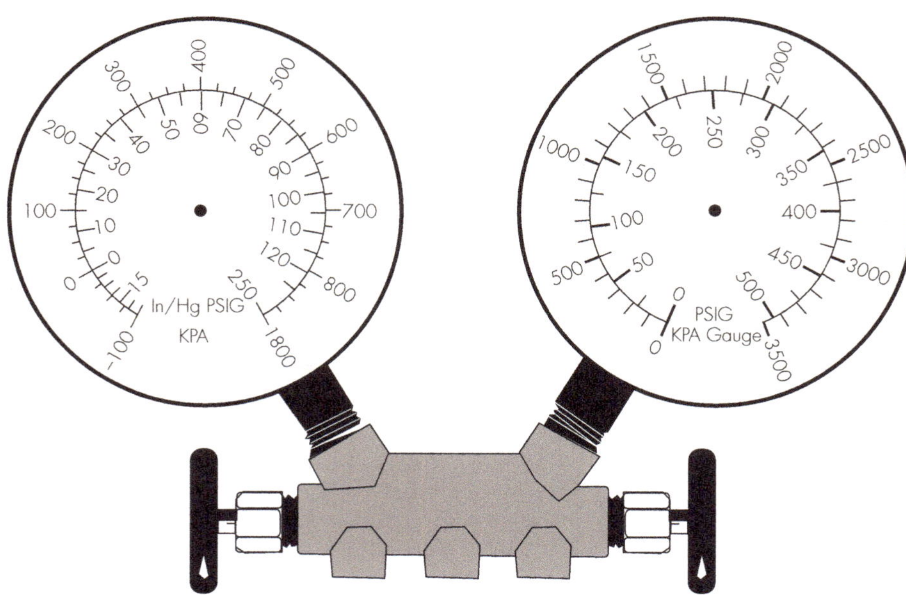

Thermostatic expansion valve – cycling clutch TX valve system

Conditions

- Ambient temperature: 35°C
- Suction (low) pressure: 40 kPa
- Head (high) pressure: 1000 kPa

Diagnosis

1 On the above manifold gauge diagram draw in the needles to indicate the pressures.

2 For this ambient temperature the normal head pressure is ______________ **kPa**

3 Normal suction pressure for this ambient temperature should be ______________ **kPa**

4 These conditions result in the evaporator cooling being **good/poor** ______________

5 The air temperature at the centre vent outlet will be ______________
–4 to 0/4 to 6/8 to 10/12 to14°C

6 This system indicates that the evaporator is **starved/flooded** ______________

7 With the above pressure readings, which component can be seen to be faulty?

System-diagnosis exercise 5

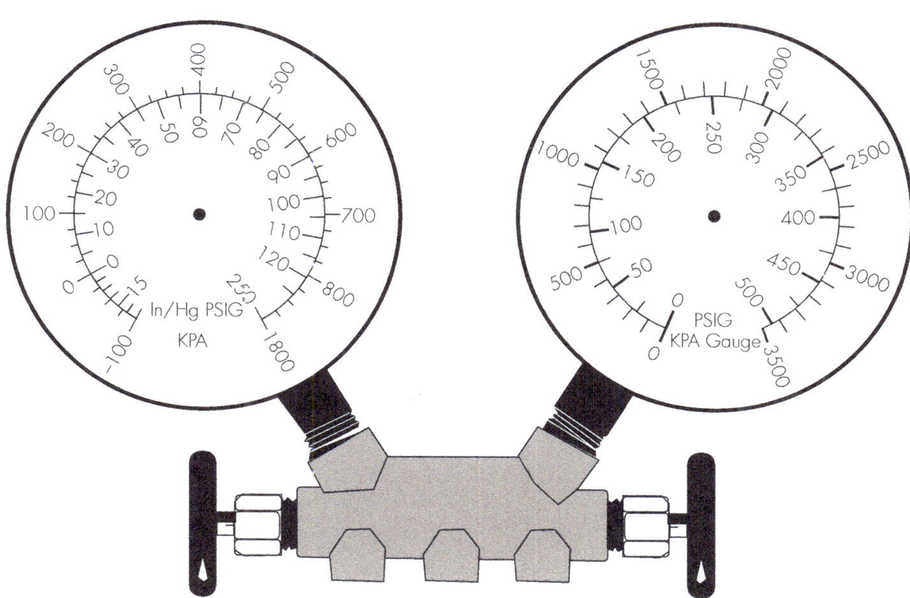

Thermostatic expansion valve – cycling clutch TX valve system

Conditions

- Ambient temperature: 35°C
- Suction (low) pressure: 380 kPa
- Head (high) pressure: 1400 kPa

Diagnosis

1 On the above manifold gauge diagram draw in the needles to indicate the pressures.

2 For this ambient temperature the normal head pressure is ____________ **kPa**

3 The suction pressure for this system is **low/normal/high** ____________

4 The head pressure for this system is **low/normal/high** ____________

5 These system pressures indicate that the evaporator is **starved/flooded** ______

6 These conditions result in the evaporator cooling being **good/poor** ______

7 With the above pressure readings, which component can be seen to be faulty?

System-diagnosis exercise 6

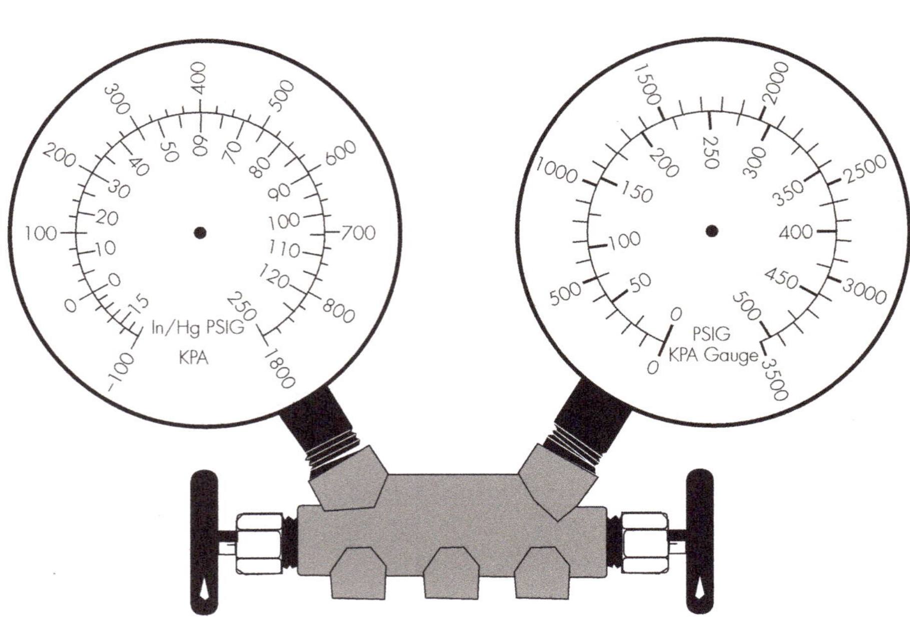

The thermostat – cycling clutch TX valve, or FOT system

Conditions

- Ambient temperature: 36°C
- Suction (low) pressure: 150 kPa
- Head (high) pressure: 1950 kPa
- Compressor cycles on longer than normal

Diagnosis

1 On the above manifold gauge diagram draw in the needles to indicate the pressures.

2 Refrigerant evaporator temperature for 150 kPa is ________________ **°C**

3 The suction pressure for this system is **low/normal/high** ________________

4 The head pressure for this system is **low/normal/high** ________________

5 These conditions result in the evaporator cooling being **good/poor** ________________

6 The air temperature at the centre vent outlet will be ________________ **–4 to 0/4 to 6/6 to 10°C**

7 This system is usually accompanied by frosting of the ________________

8 The above symptoms show that the faulty component is the ________________

9 List two faults of the component that could cause this system fault:

a ________________

b ________________

System-diagnosis exercise 7

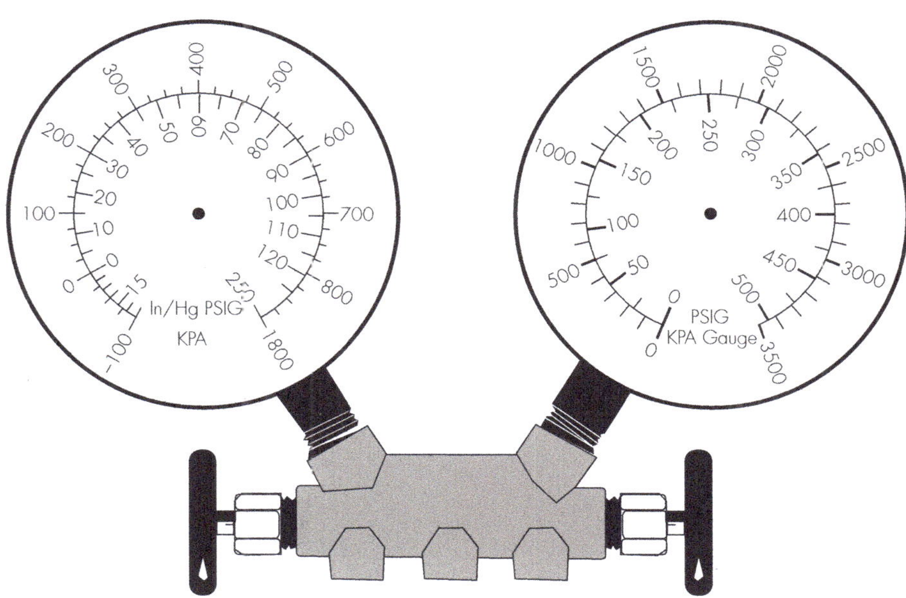

The system – cycling clutch TX valve, or FOT system

Conditions

- Ambient temperature: 32°C
- Suction (low) pressure: 710 kPa
- Head (high) pressure: 710 kPa

Diagnosis

1 On the above manifold gauge diagram draw in the needles to indicate the pressures.

2 For this ambient temperature the normal head pressure is ____________ **kPa**

3 The suction pressure for this system is **low/normal/high** ____________

4 The head pressure for this system is **low/normal/high** ____________

5 For this ambient temperature the normal suction pressure is ____________ **kPa**

6 Is the compressor working **Yes/No** ______

7 Give a reason for your answer ____________

____

8 List seven electrical faults that can cause this system fault:

a ____________

b ____________

c ____________

d ____________

e ____________

f ____________

g ____________

System-diagnosis exercise 8

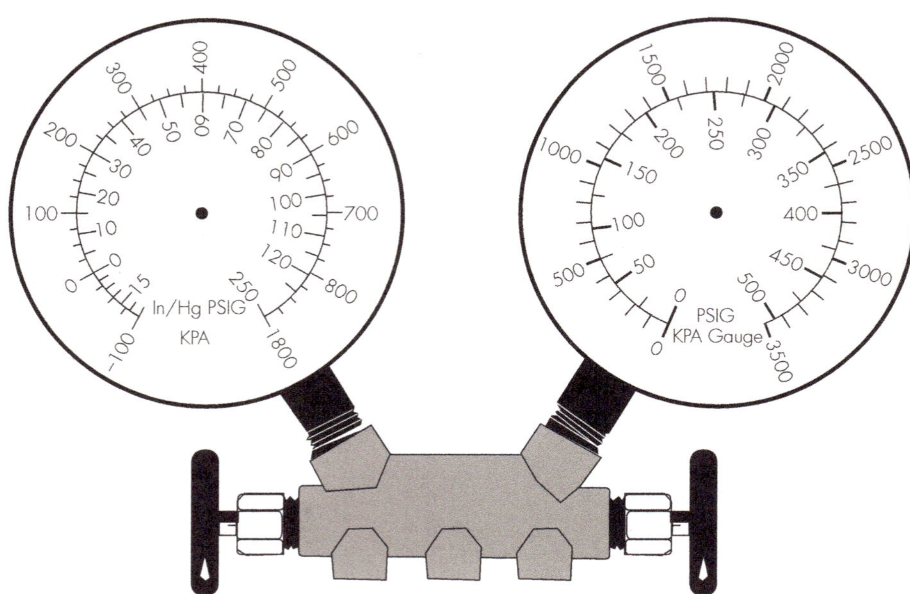

The system – cycling clutch TX valve, or FOT system

Conditions

- Ambient temperature: 35°C
- Suction (low) pressure: 210 kPa
- Head (high) pressure: 2200 kPa

Diagnosis

1 On the above manifold gauge diagram draw in the needles to indicate the pressures.

2 The suction pressure for this system is **low/normal/high** __________

3 The head pressure for this system is **low/normal/high** __________

4 The temperature at the evaporator outlet will be __________ **−2 to 0/2 to 4/4 to 8°C**

6 List two causes of this system fault:

a __________

b __________

System-diagnosis exercise 9

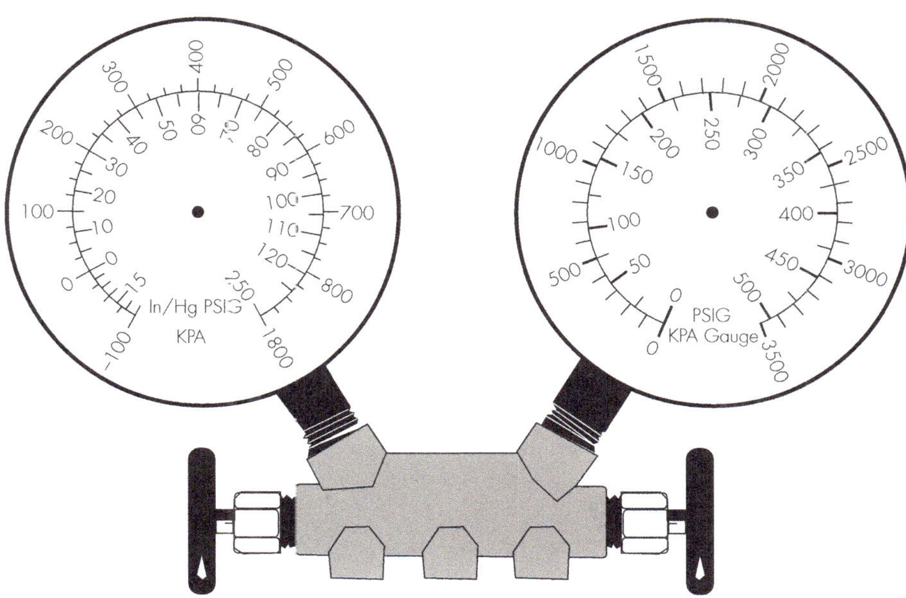

The system – cycling clutch TX valve, or FOT system

Conditions

- Ambient temperature: 36°C
- Suction (low) pressure: 310 kPa
- Head (high) pressure: 2600 kPa

Diagnosis

1 On the above manifold gauge diagram draw in the needles to indicate the pressures.

2 List six system faults that can cause excessive head pressure (poor condensing action):

a ______

b ______

c ______

d ______

e ______

f ______

System-diagnosis exercise 10

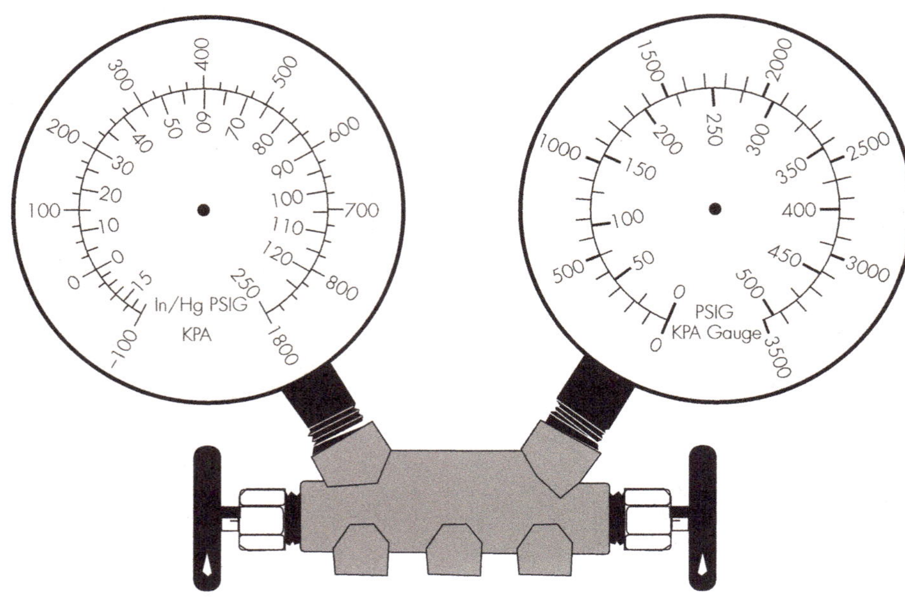

Non-cycling variable displacement receiver dryer system

Conditions

- Ambient temperature: 30°C
- Suction (low) pressure: 195 kPa
- Head (high) pressure: 1600 kPa

Diagnosis

1 On the above manifold gauge diagram to indicate the pressures.

2 For this ambient temperature the normal head pressure is ________________ **kPa**

3 Normal suction pressure for this ambient temperature should be ____________ **kPa**

4 What is the reason for the suction pressure being 195 kPa? ________________

5 Refrigerant temperature in the evaporator in this exercise is ________________ **°C**

6 The head pressure for this system is **low/normal/high** ____________

7 These conditions result in the evaporator cooling being **good/poor** __________

Give a reason for your answer __________

8 The air temperature at the centre vent outlet will be ________________

–4 to 0/4 to 6/6 to 10°C

Section 3
Service procedures

Introduction

Proper servicing and repair procedures are vital to the safe and reliable operation of the system. More importantly, the use of proper service procedures is essential if the service repair is to be undertaken safely.

The procedures given in this section are intended as a guide. Because of the many variations in system servicing, it is impossible to identify all servicing requirements in this one text.

However, in addition to those safety precautions that have already been described in this text, the following need to be observed.

Always:

- wear safety glass when working on air-conditioning systems
- operate engines in well-ventilated areas, or use exhaust extraction
- use an auxiliary air blower fan to limit the high-pressure increase (the ram air effect)
- keep all tools well away from moving parts, including the manifold and service hoses
- avoid contact with hot engine parts.

Service technicians must always exercise extreme caution and heed every established safety practice when performing any air-conditioning service procedure.

3.1 SERVICE PROCEDURE 1

Connecting the manifold and gauge set to the air-conditioning system

A manifold and gauge set showing the system low-pressure (or suction pressure – blue gauge) and high-pressure (or discharge pressure – red gauge) need to be installed whenever an air-conditioning system is being serviced. These gauges show the working pressure of the operational system for analysis – a procedure called a 'performance test'.

The refrigerant can be removed through the gauges and hoses, the system evacuated and the refrigerant put back into the system. These processes will be discussed in Service Procedures 2, 3, 4 and 5.

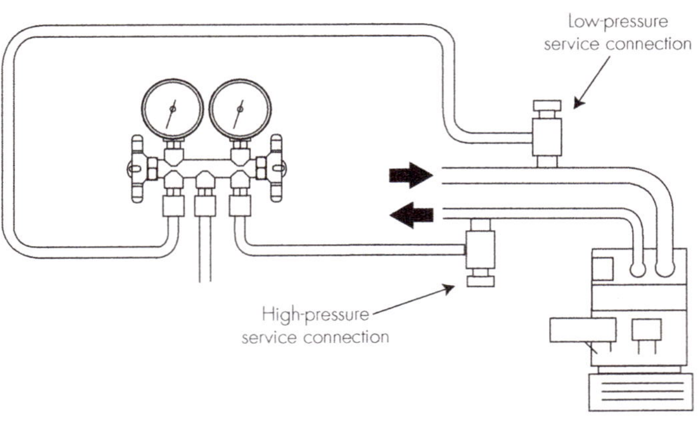

Figure SP1.1 A typical manifold and gauge connection

The process of connecting the manifold and gauge set will be covered in two parts:

- Schrader valves found on R12 systems
- quick-release valves found on R134a, R1234yf and blended refrigerants.

Recognised safety precautions must be followed when working with refrigerants.

Tools

Manifold and gauge set with three hoses:

- blue hose connected to the low-pressure/ vacuum combination gauge
- red hose connected to the high-pressure gauge
- yellow hose, which can be connected to vacuum pumps – recovery units – and bulk refrigerant sources.

All hoses must have positive shut-off valves to prevent refrigerant loss as per legislation:

- suitable trade tools and guard covers must be used
- $\frac{3}{16}$ to $\frac{1}{4}$ inch in-service valve adaptors are required to be used on some R12 connections.

Schrader valve connection R12 only

1. Remove the service valve protective cap. This cap can be made from plastic, brass or metal, and should be finger tight. Both the high- and low-pressure hose connectors can be the same size, therefore the technician must identify the correct line to connect the hose to. The low or 'suction' line is larger and cold to touch.
2. Make sure that the manifold hand valves are closed before proceeding.

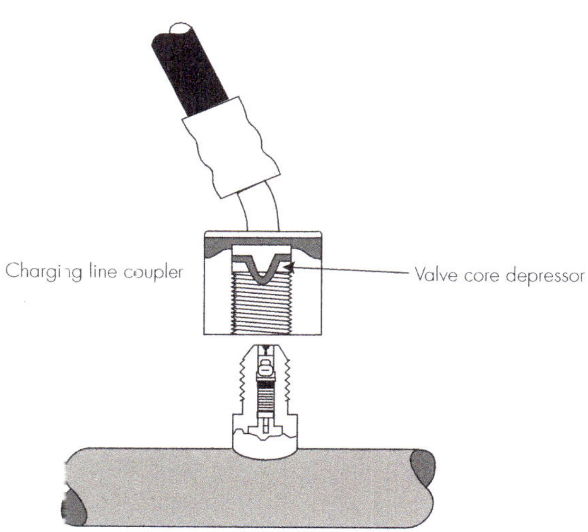

Figure SP1.2 A Schrader valve connection
Source: Atkins Carlyle Car Parts

NOTE

+ Both these service connectors are only to be finger-tightened.

Extra speed is required if the refrigerant starts to leak as the connection is made or removed. See Figure SP1.2.

3 Always replace the cap on the Schrader valve when removing the service hoses.

Quick-release coupling R134a, R1234yf and blends

Repeat steps 1 and 2 of the Schrader valve procedure.

Refer to Figure SP1.3.

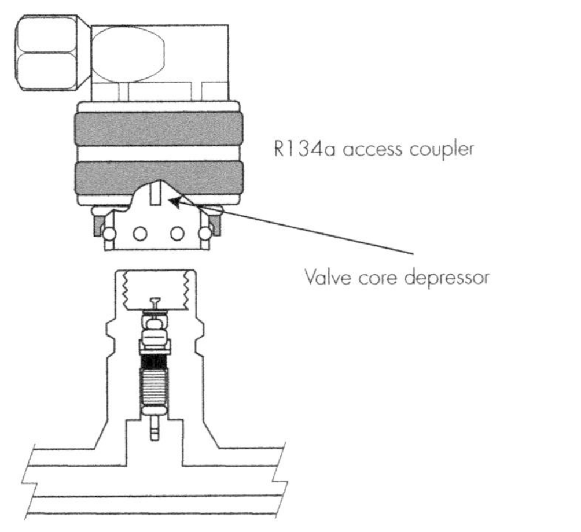

Figure SP1.3 A quick-release coupling
Source: Atkins Carlyle Car Parts

NOTE

+ The low- and high-pressure service fittings are different sizes, and cannot be fitted to the incorrect service line. Also make sure that both manifold hand valves are closed. These valves are only opened to remove and replace refrigerant in the system.

1 The quick-release connector is fitted by putting pressure over the outer ring spring force and pulling back. Place the connector over the service port, before releasing the outer ring. Lightly pull on the connector to make sure it is secured.

NOTE

+ Use two hands to connect to the system, one to fit the coupling and one to support the refrigerant line to prevent kinking and distorting the alloy piping.

2 Test the connector by gently moving it and pulling on it to see if it is locked in position.
3 Turn the hand valve on the coupling to depress the service valve.

NOTE

+ Auto coupling connections are available on gauge units hoses and do not require turning of the tap to open and close the connection.

4 The pressure gauge will usually move when the service valve is opened.
5 To remove the quick-release service fitting, first return the hand valve to the closed position.
6 Pull against the outer ring, and the connector will release.
7 Replace the valve cap.

3.2 SERVICE PROCEDURE 2

Performance testing the system

Before attempting this test, it may be advantageous to read the section on retrofitting and the various system tests that form part of the retrofit. This will ensure that you better understand the advanced performance testing procedures mentioned in this test.

Humidity affects the performance of an air-conditioning system. On a humid day the evaporator must lower the temperature of the air that enters the passenger area and also remove some of the moisture from the air.

The following performance testing procedure is a guide to the advanced diagnostic analysis required to completely check a working air-conditioning system.

The air-conditioning technician should refer to the manufacturer's service manuals for system-specific data wherever possible.

Tools

- manifold and gauge set
- pressure/temperature chart
- centre vent air temperature chart
- temperature probe
- guard covers
- eye protection
- manufacturer's specification (if possible)

Procedure

1. Connect the manifold and gauge set as shown in Service Procedure 1.
2. Make sure both manifold hand taps are closed before proceeding.
3. Start the engine and set its speed at about 1700 revolutions per minute (rpm). Allow it to warm up.
4. Place a fan unit in front of the vehicle to provide the ram air for cooling if possible.
5. Start up the air-conditioning system while the engine is warming up. This will allow it to stabilise.
6. Set the air-conditioning system controls to:
   - maximum cooling
   - recirculate air position
   - fan speed to one or two (some vehicles state high fan speed).
7. The pressures and temperatures can be taken once the compressor has cycled three or four times or the system has run for up to five minutes on a variable compressor system.

Follow the performance testing sheet on page 191.

Note the following:

- The amount of superheat can affect the condenser's ability to operate.
- The amount of temperature drop through the condenser indicates the size of the condenser and the air flow through it.
- The discharge pressure and suction pressure indicate the percentage of system charge.
- The temperature of the refrigerant in the return line shows whether or not the TX valve and evaporator are operating correctly.
- A difference of 5 to 7°C in the centre vent outlet temperature between the compressor cycle out and cycle in again indicates normal operation.

This procedure gives a complete analysis of the workings of the air-conditioning system and all its components.

Performance testing sheet

Vehicle make ______________________________

Registration number ______________________________

Year of manufacture ______________________________

Record specified temperatures and pressure

Actual ambient temperature ______________________________ °C

Specified discharge pressure ______________ kPa to ______________ kPa

Specified suction pressure ______________ kPa to ______________ kPa

Centre vent air outlet temperature ______________________________ °C

Measure and record the system information

1 Discharge pressure ______________________________ kPa

2 Discharge refrigerant temperature ______________________________ °C

3 Temperature at the condenser outlet ______________________________ °C

4 Temperature difference between item 2 & 3 ______________________________ °C

Min. temp difference should be 8 to 10°C

5 Temperature at receiver dryer inlet ______________________________ °C

6 Temperature at the receiver dryer outlet ______________________________ °C

Max. temp difference should be 2 to3°C

7 Suction pressure ______________________________ kPa

8 Suction pressure/temperature equals ______________________________ °C

9 Temperature of the suction line at the firewall ______________________________ °C

Temp at firewall should be 0 to 2°C

10 Temperature of the outlet air at the centre vent ______________________________ °C

Temp difference should be 4 to 7°C

Maximum ______________________________ °C

11 Compressor cycling in times ______________________________ sec.

12 Compressor cycling out times ______________________________ sec.

System evaluation ______________________________

3.3 SERVICE PROCEDURE 3

Purging the system and recovering the refrigerant

The removal of refrigerant from the system in order to service it is known as 'purging'. This procedure places the purged refrigerant in a recovery unit to be filtered and dried so that it can be used again or returned to the supplier for processing.

NOTE

+ Most recovery units are dedicated to one refrigerant type and should not be used to recover any other refrigerant unless specified and should only be used on those refrigerants specified.

Tools

› manifold and gauge set connected to the system, as outlined in Service Procedure 1
› recovery unit

Procedure

NOTE

+ Before starting, always measure or drain any oil already in the recovery unit.
+ Check the amount of oil in the recovery machine drain container to accurately compensate for oil removed from the system when recovery is complete.

Follow the instructions provided with the recovery unit. If they are not available, proceed as follows:

1 Connect the manifold and gauge set to the system.
2 Connect the centre (yellow) service hose of the manifold and gauge set to the refrigerant-in valve connector of the recovery unit (see Figure SP3.1).
3 Open both the low- and the high-pressure valves on the manifold.
4 Open the valves on the refrigerant bottle attached to the recovery unit.

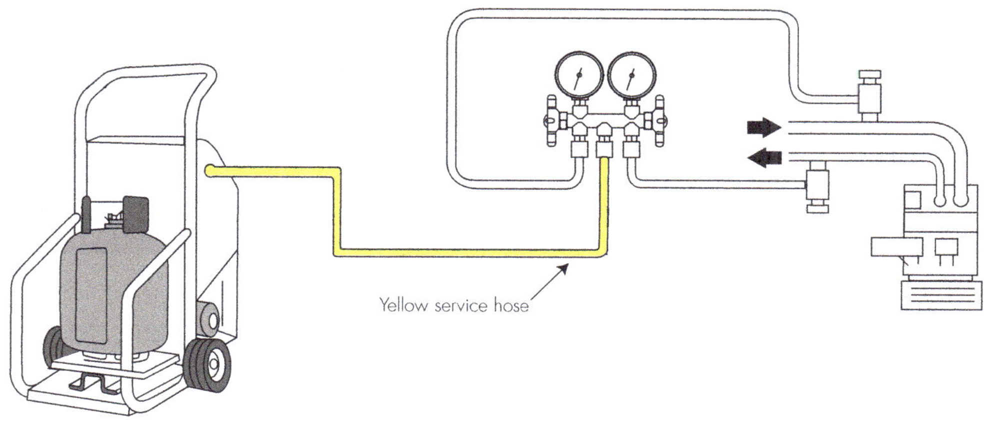

Figure SP3.1 Connecting the recovery unit to the manifold and gauge set

5 Open the valve on the recovery unit that allows the refrigerant to enter and pass through to the refrigerant bottle.
6 Switch on the recovery unit.
7 Allow the refrigerant to completely drain from the system. The low-pressure gauge will first read 0 kPa and then show a partial vacuum (–20 to –25 kPa).
8 The machine will automatically turn off once a certain partial vacuum is reached.
9 Wait to see if the low-pressure gauge rises to a pressure. The recovery unit will automatically restart to remove the last of the refrigerant.
10 Once finished, close all the opened valves and disconnect the recovery unit.
11 Drain and measure the oil removed from the system during the recovery process and stored in the machine. This amount of oil must be replaced with new oil during the service procedure.

Once the refrigerant has been recovered, the next procedure is to service the system and repair or replace any components as required. The system must then be evacuated. Most modern recovery units will automatically filter and dry the captured refrigerant as the system is being evacuated, but older machines must be manually recycled. In these machine follow the unit's instructions. However, if these are not available, proceed as follows. Note that the process can only be performed when a certain amount of refrigerant is present in the bottle, usually when the bottle is over half full.

1 Open the refrigerant cylinder valves.
2 Press the recycle switch.
3 Let the unit run for up to four hours.
4 A moisture indicator changes colour (usually from yellow to green) when the refrigerant is dry.
5 Recycling is then complete.
6 Close the cylinder valves.

3.4 SERVICE PROCEDURE 4

Evacuation of the system

The air-conditioning system must be evacuated whenever the system has been serviced or opened. This procedure removes all the moisture and air that entered the system while it was being serviced.

As pressure is lowered within a refrigerant system; the boiling point of moisture is also lowered. Placing a vacuum on the system greatly reduces the temperature at which moisture boils. This is shown in Table SP4.1, which shows that with a properly working vacuum pump it is possible to remove moisture even when the ambient air temperature is as low as 16°C.

Table SP4.1 Pressure and temperature correlation

kPa	°C
-89.6	48.9
-92.5	43.3
-94.8	37.2
-97.8	26.7
-98.8	21.1
-99.6	15.6
-100.4	10.0

Tools

- manifold and gauge set
- vacuum pump *or* a complete charging station

Procedure

The service technician should have already completed Service Procedures 1 to 3, as well as having serviced the components of the air-conditioning system. Now it is time to evacuate the system in preparation for charging refrigerant back into it.

NOTE

+ Before proceeding it is important to make sure that the low-pressure gauge is reading accurately and showing 0 kPa with no pressure on it. If it is not, disconnect it from the system and adjust it to read 0. It must be adjusted so that the exact vacuum reading is indicated. This is done by first removing the gauge face or access plug and inserting a screwdriver into the adjuster screw, then adjusting the needle to read 0 kPa.

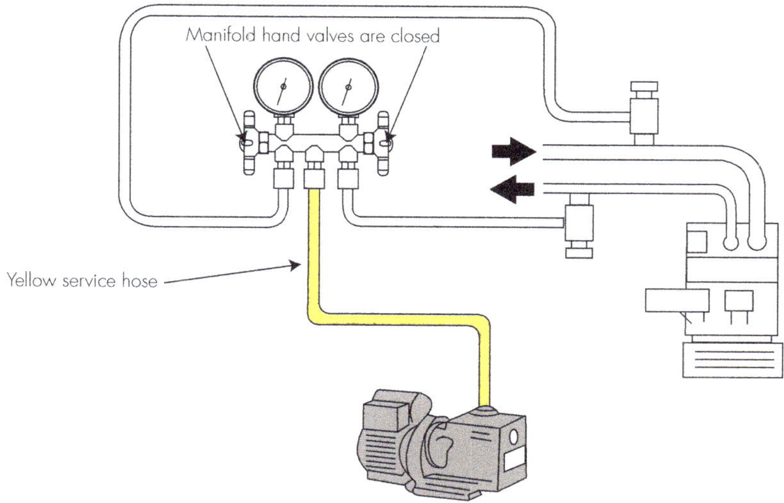

Figure SP4.1 Connecting the system to a vacuum pump

After checking that the gauges are at zero, make sure that both manifold hand valves are closed. Reconnect the hose and proceed as follows:

1 Connect the yellow centre hose of the manifold to the vacuum pump. Make sure both manifold valves are closed (see Figure SP4.1).
2 Switch on the vacuum pump.
3 Open any valves or levers on the vacuum pump.
4 Open the low-pressure manifold hand valve and observe the needle. It should move into a vacuum.
5 Observe the high-pressure gauge. It should move below 0 towards its stop. This indicates that there is no system blockage and that evacuation can continue.
6 Open the high-pressure gauge hand valve.
7 The low-pressure gauge should move to between –70 and –90 kPa within two minutes. A slight vapour may be seen coming from the pump. This is moisture being removed.
8 After four minutes the vacuum should have reached –99 to –99.6 kPa.
9 Continue operating the vacuum pump for 30 minutes, when enough moisture should have been removed to stabilise the system's vacuum level. This is the minimum evaluation time required by ARC leglislation.
10 Close both manifold hand valves and turn off the vacuum pump. Allow the unit to stand for 5 minutes.
11 If the low-pressure compound gauge loses vacuum, this indicates:
    » moisture is still in the system and further evacuation is necessary; or
    » a leak is present.
12 If no leak is indicated, continue evacuating for a further 15 minutes.

If the system has been retrofitted or converted from one using a blended refrigerant, continue evacuation for at least 30 minutes. The longer the evacuation the purer the system becomes.

13 Once the evacuation is complete, close all the valves and levers that have been opened and turn off the vacuum pump.

The system is now ready to charge with refrigerant.

Servicing the vacuum pump

See the service procedure on servicing the equipment for requirements under the legislation.

3.5 SERVICE PROCEDURE 5

System oil compensation

Oil can and will be lost whenever the system is serviced. This occurs through:

- components being replaced – e.g. receiver dryers compressors, condensers, evaporators and long runs of hose
- removal of refrigerant – oil moves around the system with the refrigerant, and therefore, as the refrigerant is removed, some oil can also be removed.

This oil *must* be replaced as most systems have less than 200 mL of oil when fully charged. It is difficult to estimate how much oil is in the system, as most modern compressors do not have oil dipstick facilities. Table SP5.1 shows the average oil compensations for each component that is replaced. The amounts described are an indication as to the components size, e.g. a small receiver dryer is 15mL, a large one is 30 mL. Manufacturer's specifications should be followed where possible.

Table SP5.1 Average oil compensation

Component	Quality
Condenser	30–40 mL
Receiver dryer	15–30 mL
Expansion valve	0 mL
Evaporator	20–40 mL
Hoses	10 mL per 600 mm
Pipes	10 mL per 600 mm
Accumulators	20–30 mL added to amount drained

Table SP5.2 Compressor oil charge table

Compressor	model	oil capacity(ml)
Delco/Frigidaire	A6	284
	R4	284
	V5V6	225
	V5V6 (after 2000)	265
	V7	265
Diesel Kiki (Zexel)	DKS12	150
	DKS13	120
	DKS15	150
	DKS16	200
	DKS17	200
	DKV14C	200
	DKV14D	200
	TM13	131
	TM15	150
	DCW	200

Compressor	model	oil capacity(ml)
Diesel Kiki (cont.)	DCW17	200
	DKY14	200
N/Denso	6D152	350
	6E171 +(A)	280
	6P80C	170
	6P127A	155
	6P127D	100
	6P134	230
	6P148	170
	10P08	100
	10P13	100
	10P15	100
Sanden	SD505	100
	SD507	135
	SD508	175
	SD510	135
	SD708+09	135
	TR90	150
	TRF105	150
Unicla	UP90	150
	UP108-110	150
	UX200	180
	UA150-170	150

This table is to be used as a guide only. The guide is not complete and all compressors available are not included in this table. Consulting OEM specifications are always recommended.

When replacing a compressor, the new compressor usually has a full system charge of oil in it; therefore this oil charge will need to be adjusted.

If the condenser, receiver dryer and evaporator are to be replaced with new components, leave the full charge of oil in the compressor. An adjustment will need to be made if the condenser and evaporator are to be left in the system. This could mean the removal of some 50 mL of oil from the compressor.

This is not an accurate method of determining the correct charge rate of system oil, but will place enough oil in the system to prevent damage.

If the system was flushed before the new compressor was placed in the system, the full oil charge can be left in the compressor. This is one of the advantages of flushing the system. The correct oil charge is now easily determined, while new pure oil replaces the older contaminated oils.

Oil can either be placed in the component being replaced, injected into the system at the beginning of the evacuation or flushed in with the charging of the system. The method chosen depends on the equipment available in the workshop.

3.6 SERVICE PROCEDURE 6

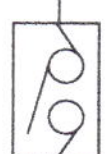

Charging the system

Only a liquid source should be used when a system is being charged. There are several ways of introducing the refrigerant for charging. These include:

- a charging station with a 4 or 5 kg *Dial-A-Charge* cylinder – do not use on a Zeotrope refrigerant
- a 25 kg cylinder with scales and gauges
- a fully-integrated electronic servicing station.

Some cylinders with one tap can be marked either 'vapour' or 'liquid'. A cylinder marked 'vapour' needs to be turned upside-down to allow the liquid to flow.

Two different ways of carrying out the procedures will be described here.

Charging from a *Dial-A-Charge* unit

Equipment

- manifold and gauge set
- *Dial-A-Charge* cylinder

Filling the unit

If the top hand tap of the *Dial-A-Charge* unit is allowed to vent while it is being filled, about 1 kg of refrigerant will be lost to the atmosphere. However, this refrigerant is captured, stored, filtered and re-used, as per legislation requirements, thus lowering the cost of the procedure. See Figure SP6.1.

1. Connect a yellow hose from the bulk refrigerant cylinder to one of the bottom hand taps on the *Dial-A-Charge* unit as shown in Figure SP6.2 (hose A). Open the bulk cylinder and the *Dial-A-Charge* lower hand taps. This allows refrigerant to flow into the unit.
2. Connect another yellow service hose from the top hand tap or vent to the recovery unit refrigerant inlet connector or tap as shown by hose B.
3. *Slightly* open the top hand tap or vent to allow refrigerant to escape slowly and operate the recovery unit according to the instructions or Service Procedure 3.
4. When the *Dial-A-Charge* unit is nearly full, close the top valve or vent and then turn off the refrigerant supply. (Close all taps.)
5. Finally turn off the recovery machine and remove the yellow service hoses.

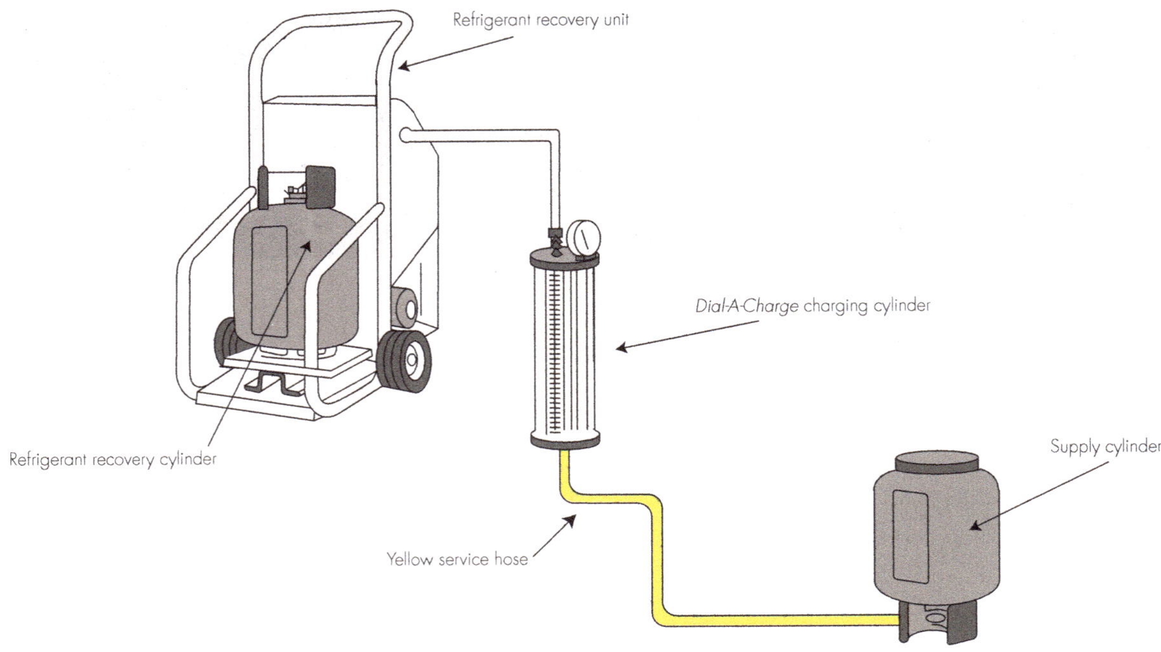

Figure SP6.1 Charging from a *Dial-A-Charge* unit

Setting the *Dial-A-Charge* unit

There is a pressure gauge that shows the pressure on top of the *Dial-A-Charge* cylinder. This, therefore, shows the temperature of the refrigerant, which should be the ambient air temperature near the cylinder. The pressure is usually shown in kPa and bar.

The outer surface or sleeve of the cylinder can be moved to adjust the refrigerant scale to the given pressure shown on the top pressure gauge. This refrigerant scale must cover the refrigerant tube.

As an example, if the pressure gauge is showing 650 kPa or 6.5 bar, move the outer scale around until the 6.5 or 650 kPa scale lines up with the refrigerant tube. The correct scale for that pressure is now shown and ready for use.

Move the rubber ring down the outer sleeve past the liquid level to indicate the amount of refrigerant that needs to be charged (e.g. 500 grams). The system can now be charged.

Charging from a bulk source

Equipment

- manifold and gauge set
- bulk refrigerant
- scales
- *or* an electronic charging station

Place the bulk source on the scale set, either upright or upside-down. If the bulk container has two hand taps (one marked 'vapour' and the other marked 'liquid') leave the cylinder in the upright position and connect the yellow manifold hose to the liquid hand tap on the cylinder. If it has only one tap and it is marked 'vapour', turn it upside-down. This will supply liquid to the system.

Set the scales by first weighing the bottle and then setting to zero. The amount of refrigerant to be charged into the system can now be set off the scale (e.g. 500 grams).

Charging the system

1. Obtain and note the system refrigerant charge from Table SP6.1.
2. Connect the refrigerant source to the manifold by the centre yellow hose A, as shown in Figure SP6.2.
3. Open the hand taps on either the *Dial-A-Charge* cylinder as indicated by B, or the liquid hand tap on the bulk cylinder.
4. Slowly open the low-pressure (blue) hand valve (C) on the manifold.

NOTE

- Control the pressure of the refrigerant entering the system by *slightly* opening and then adjusting the hand valve. Limit the pressure to a maximum of 300 kPa. This will prevent liquid entering the low-pressure side of the compressor and causing damage.

5. Continue to charge the system until both gauges rise to 300kPa. Open the low-pressure tap and allow extra refrigerant to enter the system until the refrigerant stops

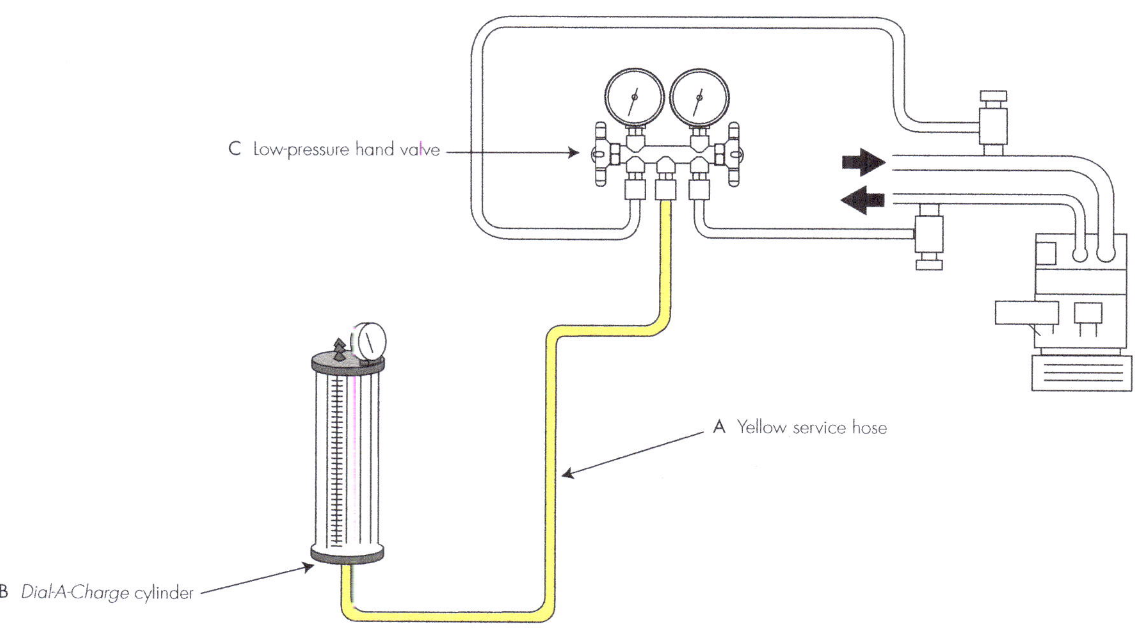

Figure SP6.2 Charging the system

flowing. Pressure in the cylinder is the same as pressure in the system.

6 Close the low-pressure hand valve.

NOTE

+ The following are not required if an integrated recovery charging station is used as the unit has its own pumping system and will place the full charge in the system.

7 Start the vehicle and run the air-conditioning unit. The pressure on the suction side of the system will drop, allowing extra refrigerant to be drawn into the system. *Only* open the low-pressure tap during this operation.

8 Only allow refrigerant to enter the system when the compressor is engaged and working. If the compressor cycles off, close the low-pressure hand valve and wait until the compressor cycles back in.

9 Continue charging the system until the recommended amount of refrigerant has been charged into the system.

10 Close all hand taps.

11 Performance test the system.

This capacity guide is not complete and is used only as a guide. It is recommended that you source OEM specifications. The specifications for current vehicles are not always available for publication.

Table SP6.1 R134a refrigerant capacity guide – all capacities are +/– 50 grams

Vehicle	Year model	Capacity (g)
Alfa Romeo		
145	94–97	700
146	96–00	700
147	04–06	700
155	94–97	700
Alfa Romeo (cont.)		
156	97–01	650
156	02–06	500
159	05–08	550
166	98–01	700
166	May 01–07	650
Alfetta	All	460
Giulietta	All	650
GTV	98–03	700
GT Coupe	03–06	600
GT Coupe V6	All	650
Spider	98–03	700
Spider	Sept 03–04	575
Audi		
A3	97–02	750
A3	03–07	525
A4	97–01	725
A4 B6	01–06	500
A6 C5	99–01	675
A6 4f	04–06	530
80	93–95	650
100	R134a models only	600
200	R134a models only	600
BMW		
1 Series	04–06	500
3 Series E92	06–09	600
3 Series E46	98–06	725
3 Series E91	2005 onward	600
323i	93–99	900
325i	96–01	675
328i	96–00	925
5 Series E60/61	03–09	790
6 Series E63/64	04–07	800
7 Series E38	98–01	670
7 Series E65	02–05	800
X3	03–07	800
X5	99–06	440

Vehicle	Year model	Capacity (g)
BMW (cont.)		
Z3	96–02	450
Z8	00–04	850
Chrysler/Dodge		
300C	04–06	750
Crossfire	03–06	850
Neon	96–98	700
Neon	99–01	650
Neon	01–06	785
PT Cruiser	01–02	675
PT Cruiser	03–05	510
Voyager	Single unit	850
Voyager	Dual unit	1350
Citroen		
Berlingo	99–03	575
C2	04–06	600
C3	02–06	625
C4	2000 onwarc	450
C5	01–04	625
C6	2000 onwarc	625
C6	00 on diesel	525
Xantia SDV16	94–98	875
Xantia	98–01	825
XM	94–00	1000
Xsara	98–00	975
Xsara	00–05	565
Daewoo		
Cielo	R134a models cnly	730
Espero	R134a models cnly	730
Kalos	03–04	600
Lacetti	03–04	720
Lanos	03–07	720
Nubira	Before 97	750
Nubira	97–00	825
Leganza	97–02	830
Matiz	99–01	550
Matiz	02–08	450

Vehicle	Year model	Capacity (g)
Daewoo (cont.)		
Musso	00–03	700
Tacuma	00–04	550
Daihatsu		
Applause	93–97	550
Applause	97–99	450
Charade	93–03	650
Charade	03–05	300
Feroza	93–97	500
Mira	95–97	500
Pyzar	97–00	650
Rocky	93–99	550
Sirion	98–05	380
Terios	97–01	600
Ford		
Courier	93–97	500
Courier	97–02	550
Courier	02–06	850
Explorer	96–01	650
Explorer	01–06	900
Falcon	93–94	675
Falcon	EF–EL	750
Falcon	AU	850
Falcon	02–05	650
Fiesta	2002 onward	650
Focus	02–06	750
Laser	99–02	625
Mondeo	96–00	765
Mondeo	2000 onward	750
Probe	94–98	750
Territory	04–02	650
Holden		
Astra	Before 2000	700
Astra	01–04	600
Astra	04–06	500
Barina	01–04	640
Barina	04–10	500

Vehicle	Year model	Capacity (g)
Holden (cont.)		
Captiva	00–05	650
Combo	01–06	650
Commodore	VP–VZ	800
Cruze	02–06	375
Jackaroo	93–03	750
Rodeo	97–03	650
Rodeo	03–06	750
Vectra	03–06	730
Viva	03–06	680
Honda		
Accord	94–97	600
Accord	98–02	625
Accord	03–06	525
Civic	94-00	525
Civic	00–06	525
CRV	97–01	625
CRV	02–06	500
Integra	94–01	625
Integr	01–06	525
Jazz	02–06	425
Legend	96–04	725
MDX	03–06	700
Odyssey	00–04	975
Odyssey	04–06	725
Prelude	94–96	625
Prelude	97–00	725
Hummer	03–04	800
Hyundai		
i30	2008 onward	500
Accent	00–03	600
Accent	03–06	550
Elantra	96–02	700
Excel	94–00	600
Getz	02–06	500
Santa Fe	00–06	550
Sonata	01–06	700

Vehicle	Year model	Capacity (g)
Hyundai (cont.)		
Terracan	01–06	850
Tucson	00–06	550
Jaguar		
S type	00–02	775
S type	03–06	675
X type	01–06	775
XK8	96–03	600
XKR	02–06	650
Jeep		
Cherokee	95–96	800
Cherokee	01–06	750
Wrangler	96–06	500
KIA		
Carnival	00–06	800
Cerato	04–06	500
Credos	98–01	700
Magentis	2000 onward	500
Mentor	96–00	500
Optima	01–06	675
Rio	98–01	465
Rio	01–06	575
Sorento	03–06	575
Spectra	01–03	700
Sportage	97–03	700
Sportage	04–06	525
Lada		
Niva	94–98	650
Land Rover		
Defender	95–97	800
Discovery	98–06	750
Freelander	97–00	725
Freelander	00–06	550
Lexus		
ES300	98–01	650
ES300	02–06	800
CS300	97–04	600

Vehicle	Year model	Capacity (g)
Lexus (cont.)		
GS400	98–00	550
GS400	01–03	550
GX470	03–04	550
GX400 dual	03–04	700
IS200	99–05	550
IS300	99–05	600
LS400	96–00	650
LS430	00–06	550
LX470	98–06	1050
RX330	03–05	550
SC430	01–05	475
Mazda		
121	96–02	600
2	02–06	475
3	02–06	625
323	98–03	625
6	02–06	400
626	98–00	650
929	96–97	700
B2600	2003 only	750
Bravo	99–02	550
Bravo	02–06	850
BT50	2007 only	450
MPV	01–03	550
MX5	94–97	650
MX5	98–01	450
Premacy	93–97	720
Premacy	99–02	525
Protégé	95–03	500
RX8	02–06	430
Tribute	01–06	850
Mercedes-Benz		
A class	98–05	600
C class	94–97	850
C class	01–04	700
C class	04–06	850

•••

Vehicle	Year model	Capacity (g)
Mercedes-Benz (cont.)		
C class estate	93–01	850
CL500	98–03	1050
CL 600	98–03	1050
CLK	97–03	850
CLK	03–06	750
CLK 320-430	98–03	750
CS 1200	95–00	1200
E 320	94–00	850
E 320	01–03	900
E 420	94–00	850
E 430	98–00	850
E class	96–02	1000
E class	02–06	950
ML 320–430	98–03	700
S class	91–99	1150
S 500	00–01	1050
S 500	01–03	850
SL class	99–06	900
SLK class	00–04	850
SLK 230	98–03	750
Sprinter	98–06	800
Vito	98–04	880
Vito	04–06	550
MG		
F	97–02	625
TZ	02–05	450
Mini		
All	02–08	425
Mitsubishi		
380	05–06	450
380	05–06	450
3000GT	94–00	650
Colt	04–06	550
Express	97–05	725
Galant	93–97	750
Grandis	04–06	560

•••

Vehicle	Year model	Capacity (g)
Mitsubishi (cont.)		
L300	00–06	625
Lancer	97–02	575
Lancer	03–06	550
Magna	94–96	775
Magna	96–98	675
Magna	99–00	670
Magna	00–03	650
Magna	03–05	675
Mirage	97–02	500
Nimbus	93–98	700
Outlander	03–06	550
Pajero iQ	99–02	575
Pajero	93–00	750
Pajero	00–06	550
Starwagon	93–02	700
Triton	94–97	700
Triton	97–04	575
Triton	2005 only	725
Nissan		
Navara	97–06	600
Pathfinder	94–05	650
Pathfinder	05–07	650
Patrol GU	97–06	800
Patrol	00–07	775
Pulsar	95–06	550
Terrano	97–99	750
Tida	06–11	450
X trail	01–07	500
Peugeot		
206	99–06	650
306	94–97	900
306	97–02	725
307	01–06	585
405	94–97	675
406	96–99	750
407	04–06	626

Vehicle	Year model	Capacity (g)
Peugeot (cont.)		
605	94–97	885
607	01–06	626
Porsche		
911	94–98	840
911	98–07	800
928	93–95	1000
Boxter	97–07	750
Cayenne	03–06	700
Proton		
Persona	95–05	550
Satria	97–05	500
Renault		
Cilio	01–06	650
Laguna	95–96	800
Laguna	02–06	650
Magane	01–04	760
Magane x84	03–06	550
Senic	001–02	750
Senic	2 litre	680
Senic	05–06	550
Trafic	04–06	700
Rover		
75	01–05	450
Range Rover	95–02	1250
Range Rover	02–06	520
SAAB		
9-3	02–06	750
9-3(440)	02–06	675
9-5	01–06	875
900	93–98	800
9000	93–98	775
Skoda		
Octavia	04–08	500
Subaru		
Forester	94–96	750
Forester	97–01	650

Vehicle	Year model	Capacity (g)
Subaru (cont.)		
Forester	02–04	600
Forester	04–08	550
Impressa	97–02	550
Impressa	03–08	460
Liberty	94–99	650
Liberty	98–03	475
Liberty	03–06	400
Liberty	04–08	400
Outback	94–99	650
Outback	99–03	450
Outback	03–08	400
Suzuki		
Baleno	96–01	600
Grant Vitara	98–05	475
Ignia	00–05	360
Jimny	98–06	550
Liana	01–06	500
Swift	93–01	475
Vitara	99–04	475
Toyota		
4 Runner	93–96	600
Avalon	00–11	500
Avensis	01–03	485
Camry	97–02	800
Camry	02–06	580
Camry	06–11	500
Celica	94–99	650
Celica	99–05	450
Corolla	94–01	650
Corolla	01–06	500
Hiace	94–06	700
Hilux	93–97	750
Hilux	97–01	550
Kluger	05–06	650
Landcruiser	93–98	850
Landcruiser	98–06	675

Vehicle	Year model	Capacity (g)
Toyota (cont.)		
MR2	93–06	475
Prado	96–02	750
Prado	03–06	650
Prius	00–07	450
Rav 4	94–00	700
Rav 4	00–06	525
Tarago	93–00	900
Tarago	00–06	700
Yaris	05–10	400
Volkswagen		
Beetle	00–07	700
Bora	99–06	775
Cabrio	95–02	650
Caddy	2005 only	500
Golf	98–04	750
Golf	04–06	525
Jetta	98–06	1150
Passat	95–97	1150
Passat	97–00	650
Passat GP	01–05	600
Passat GPW8	03–05	500
Touareg	03–06	725
Volvo		
C70	98–01	700
C70	02–04	800
S40	97–04	850
S40m	04–06	550
S60	00–06	1000
S70	97–00	800
S80	98–05	1000
V40	97–04	850
V50	04–06	550
V70	97–00	800
V70	00–06	1000
V90	94–98	900
XC90	03–06	1000

3.7 SERVICE PROCEDURE 7

Leak testing the system

Locating refrigerant leaks can be difficult, but there are two methods of locating the leak:

- using electronic leak detectors as specified within the ARC legislation
- using dye and fluorescent dye leak detection – used as an alternative if the leak cannot be found using the electronic detector.

Each type of leak detector will be discussed in this section.

Electronic leak detectors

1 The halogen electronic leak detector

These can only be used for a refrigerant containing chlorine – so they can only be used on R12.

Halogen electronic leak detectors, illustrated in Figure SP7.1, are less sensitive than dye and fluorescent dye leak detectors.

The halogen detectors are small, portable, relatively inexpensive and can find refrigerant leaks as small as 7 grams per year.

NOTE

+ Halogen detectors require their sensing tips to be replaced every six to nine months if the continuing accuracy of the unit is to be maintained.

The unit's battery source must also be fully charged in order to ensure its accuracy.

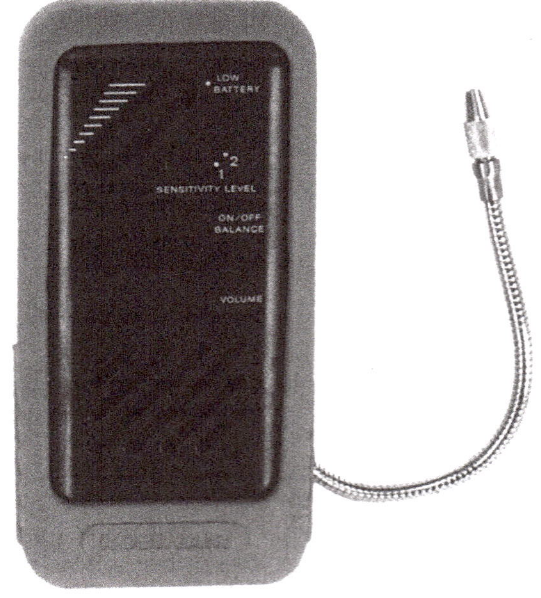

Figure SP7.1 A halogen electronic leak detector
Source: Atkins Carlyle Car Parts

2 The electrochemical heated diode detector

These units use refrigerant-selective technology and can locate leaks as small as 5 grams per year in all refrigerant types.

Using electronic leak detectors

When leak testing a system with electronic leak detectors, all joints, connectors and fittings must be free of old oil stains, as the oil can contain residue refrigerant that could give a false reading.

It is important that a pressure of at least 350 kPa must be present in the system. If this pressure is not there a serious leak is present and dry nitrogen or LP gas should be used with the appropriate leak detector.

Although the operating procedures for the various brands of electronic leak detectors may vary, the following steps should be common to all of them.

1. Switch the unit on near the area in which it is to be used. If the detector is to be operated inside the vehicle, it should be turned on inside the vehicle away from any suspected leak. This will eliminate false readings from the dash and carpet fumes.
2. Allow the unit to warm up, which can take up to 10 seconds.
3. If the unit has a sensitivity control, this must be adjusted to give a regular ticking noise (one to two times per second).
4. Move the unit's probe *under* all hoses, joints and components at a speed of 25 to 50 mm per second.
5. The ticking or lights on the unit will react when a leak is detected, thus allowing you to identify the area. Check the complete system as there can be more than one leak.
6. Never leave the probe near a leak for any length of time, as this can affect the sensitivity of the unit. (The sensing tip may be poisoned and need replacement.) Water and moisture also affect the sensing tip operation.

Fully charge the system after all leaks have been located and repaired, the receiver dryer replaced and the system evacuated. Finally, check one last time for any leaks before returning the vehicle to the owner.

Dye and fluorescent leak detectors

The coloured dye solution is introduced into the system during the servicing process. There are two types: one that mixes with the oil and one that mixes with the refrigerant and flows around the system. As refrigerant escapes, the oil also escapes, thus allowing leaks to be identified. Leaks as small as 3 grams per year can be detected by this leak detection process.

Once the dye has been introduced into the system, it remains there until the oil or refrigerant is flushed out.

Some dye solutions are refrigerant-specific and must be used only with that specified refrigerant. However, modern dye solutions can be used with all refrigerants. The instructions must be read carefully and strictly followed.

Dye solutions do not affect the operation of the air-conditioning system in any way when used as a once-only treatment, but over-use or multi-dosing of the system can harm the system components, especially the condenser. The refrigerant can be recovered and recycled without any problems, but lost oil must be replaced when servicing.

Preparing to use the dye

The manifold and gauge set should be connected and there should be no refrigerant in the system. There are three alternative methods for introducing dye to the system:

1 *A charge adaptor infuser or injector* (see Figure SP7.2):
   - change the system valves as part of this process
   - connect the dye charge adaptor or infuser into the low-pressure charging line
   - measure the required amount of dye into the adaptor or infuser
   - charge the system, following the instructions given in Service Procedure 5
   - the dye will be injected into the system with the refrigerant.

2 *A single charge (one shot) disposable tube*:
   - The service valve is removed from the service port and the dye charge injected into the port. This is done at the beginning of the evacuation process, as the vacuum helps the dye injection process.
   - The new service valve is now installed and the manifold and gauge set fitted. The system can now continue being evacuated and charged, following the instructions in Service Procedures 4 and 5.

3 *Pre-mixed refrigerant/dye* – the dye is mixed with the refrigerant at manufacture

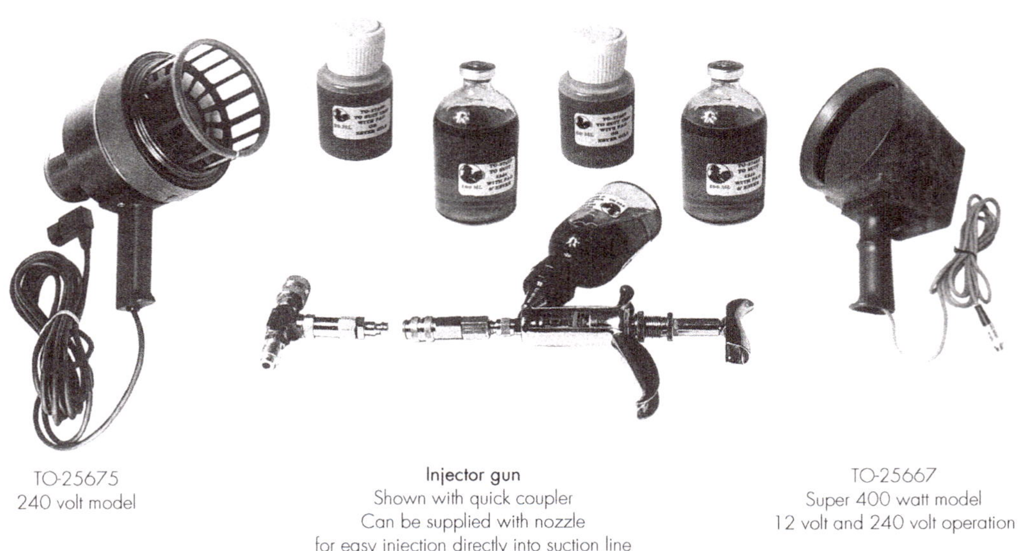

Figure SP7.2 A Glo-Leak electronic leak detector
Source: Atkins Carlyle Car Parts

and is charged into the system normally with the refrigerant. No extra dye should be injected when using refrigerant with dye already in it.

Using the dye leak detection system

With the dye in the system, operate the air-conditioning system for 10 minutes (large leak) or up to a day (small leaks).

Use an ultraviolet light (black light) and proceed with the following:

1 Shine the light at every joint, hose, seal and component.
2 Check for signs of the dye.
3 If there are no signs of a leak, but one is suspected, have the vehicle return after 24 or 48 hours to re-check for leaks.

Leaks at the evaporator can be located by allowing the system to run normally for 24 to 48 hours and then checking for leaks at the water drainpipe.

3.8 SERVICE PROCEDURE 8

Testing and adjusting the thermostatic switch

Cycling clutch systems

The thermostat controls the temperature of the refrigerant in the evaporator by cycling the compressor.

It supplies voltage and current to the compressor clutch, usually via relays and pressure switches through the ECU. The thermostat may need to be adjusted occasionally as a result of wear, and to suit local altitude conditions.

If the thermostat is electronic (when it is known as a 'thermistor') there may be no provision for adjustment.

Variable capacity compressors

On non-cycling compressors, the thermistor relays evaporator temperature to the control valve (mechanically) or to the ECU that controls the compressor's variable operation.

Testing thermostat operation

The air-conditioning system must be working normally when the thermostatic switch is being tested.

1 Operate the system for five minutes under performance testing conditions.
2 Check the ambient air temperature.
3 Use a temperature probe in the centre vent to measure the outlet air temperature.
4 Check the pressure/temperature chart on page 170 for the correct centre outlet air temperature for the measured ambient air temperature.
5 The thermostat should cut out in the operating range for that ambient temperature.
6 Finally, the outlet air should rise by between 5 and 7°C before the thermostat cuts back in.

For electronic controlled systems a scan tool is required and the operating evaporator temperature needs to be checked if it is operating within the correct specifications

Adjusting the thermostatic switch (mechanically operating units)

There are usually two adjusting screws on the thermostat:

- temperature adjustment (evaporator temperature – it cycles the compressor off)
- sensitivity adjustment (the temperature difference between compressor cycle on and cycle off). This sensitivity adjustment is difficult and should not be attempted unless absolutely necessary.

Adjusting the thermostat

The system must have no other fault if this adjustment is to be made.

1 Locate the thermostatic switch (see Figure SP8.1). It is usually under the dash on the passenger side near the evaporator.
2 Locate the temperature adjusting screw (see Figure SP8.2) which is usually under the plastic end cover and covered with red paint.
3 The adjusting screw will have directional arrows to indicate which way to turn it to increase or decrease the temperature setting.

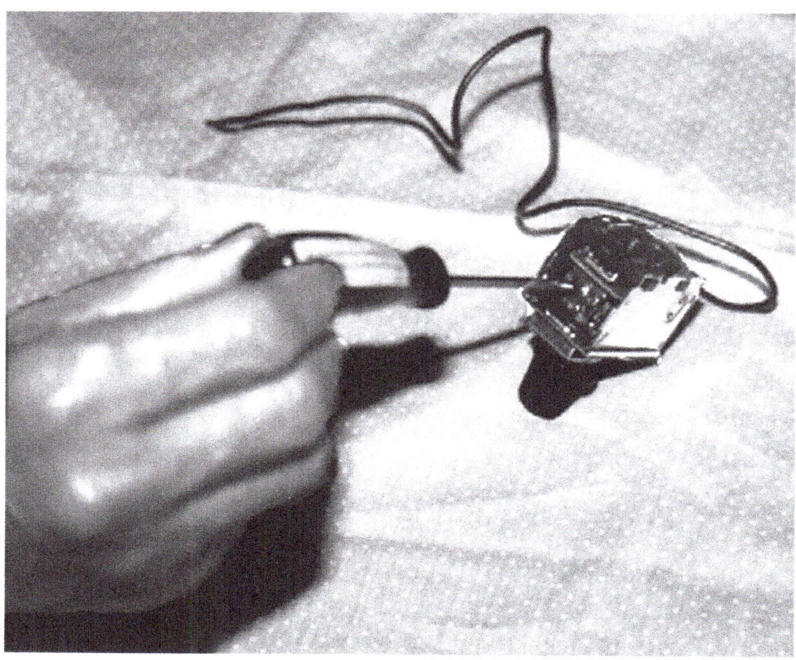

Figure SP8.1 The thermostatic switch

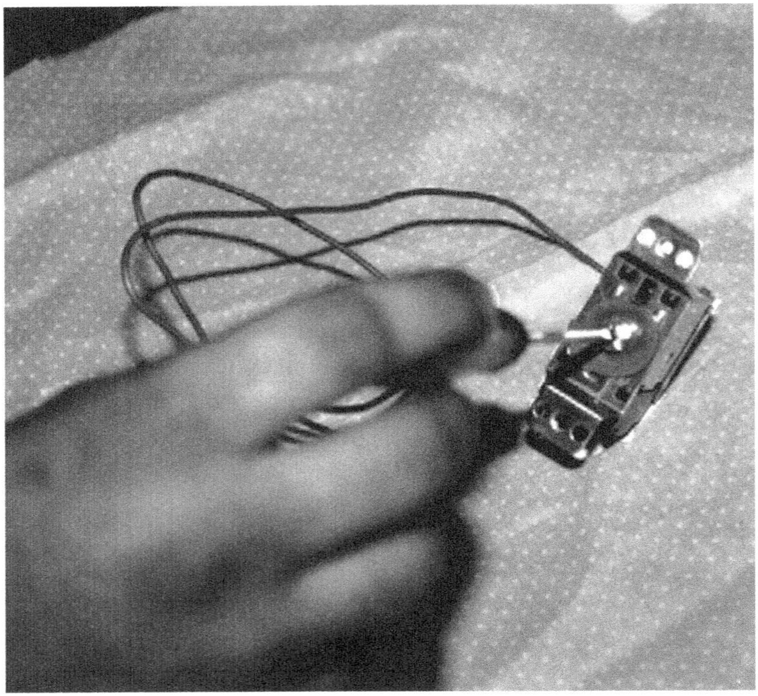

Figure SP8.2 The temperature adjusting screw

4 Adjust the screw by a quarter of a turn in the correct direction.
5 Before attempting any further adjustment, let the compressor cycle three times in order to check the cycling temperatures.
6 The thermostatic switch must be replaced if the correct cycle temperatures and cycle times cannot be achieved.
7 Replace any covers and return the system to normal operation.

Caution

- If the switch must be moved to enable adjustment, take care not to disturb the pipe in the evaporator as this could result in damage.
- The temperature sensitivity screw only needs a slight movement to send it out of adjustment and it is difficult to return to its original setting.
- Do not set the cycle temperature too low, as this could cause evaporator freezing and system failure.

3.9 SERVICE PROCEDURE 9

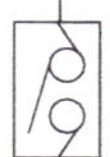

Testing the expansion valve – TX valve or FOT

Not all expansion valves can be properly tested in the vehicle. This service procedure applies to those expansion valves that can be tested in it.

The air-conditioning system must be operational and fully charged before proceeding. Connect the manifold and gauge set and run the system at 1500 to 1700 rpm. Operate the unit for 5 minutes.

Testing the expansion valve – TX valve

LOW (suction) pressure reading

As the compressor cycles in the low-pressure gauge should fluctuate, this indicates that the TX valve is opening and closing (modulating). It is sensing the refrigerant's temperature. If this does not occur, continue testing the unit with the following steps:

- Remove the valve-sensing bulb from the suction line and hold it in the hand or place hot water over it. Observe the low-pressure gauge reading. If it rises, the bulb was not correctly placed. Clean and reposition the bulb correctly, in its cradle on the suction line, using the correct clip. Do not use cable ties or tape. Finally, insulate the assembly using the correct bitumen tape.
- If the pressure is still not correct then place a hot rag, or pour hot water, over the valve body and observe the gauge reading.
- If the pressure rises to normal or near normal, moisture is present in the system. Evacuate and recharge the system and retest.
- If the above steps have not fixed the fault, replace or remove and service the unit. The receiver dryer must also be replaced.

HIGH low-pressure reading

- If the low pressure is higher than specifications it may be jammed and will not fluctuate when the compressor is operating.

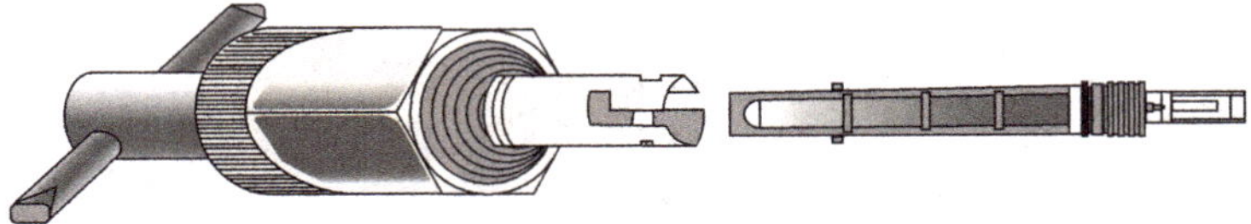

Figure SP9.1 Inserting the removal tool in a fixed orifice tube

- Check if the system has been retrofitted, as the expansion valve may need replacing if its flow rate is too high.

Testing the fixed orifice tube (FOT)

Fixed orifice tubes are not all the same sizes.

Testing the FOT is reasonably simple as there are only two basic tests – one for moisture and one for dirt blockage.

Prepare the system as previously described. The low-pressure switch may need to be bypassed in order to maintain system operation.

LOW low-pressure reading

- Place a hot rag, or pour hot water, around the tube pipe and observe the low-pressure reading.
- If it rises to normal or near normal, moisture is present in the system. Evacuate the system and change the accumulator.
- If the pressure does not rise, the FOT screen is clogged and must be removed, cleaned or replaced.

HIGH low-pressure reading

If the flow rate of the FOT is too high the restrictor has moved and the FOT must be replaced

Removing the FOT

Once the refrigerant has been purged from the system, remove the liquid line connection into the evaporator.

- Pour a small amount of clean system oil into the pipe to help with the removal of the FOT.
- Insert the removal tool as shown in Figure SP9.1. Fixed orifice tubes can be very tight and may not be able to be removed with long-nosed pliers.
- Turn the tee bar to lock the tube in place.
- Turn the outer sleeve to pull the tube out of the pipe.
- If the tube breaks, the extraction tool must be used to remove the stuck broken parts.

Refitting the new FOT

Coat the new tube with system refrigerant oil and place it into the liquid line pipe until it is fully seated. Replace the accumulator and disturbed O-rings.

Evacuate and charge the system. Finally, performance test the unit.

3.10 SERVICE PROCEDURE 10

Servicing the Sanden SD compressor

Oil level measurement (in vehicle)

The oil level in the compressor should be checked when a system component has been replaced, when an oil leak is suspected or when it is specified as a diagnostic procedure.

1 Run the compressor for 10 minutes with the engine at idle.
2 Recover all refrigerant from the system. Do this slowly so as not to lose any oil.
3 Determine the mounting angle of the compressor from the horizontal (i.e. oil plug or adaptor on top). If access to the compressor permits, this is easiest done by using a machinist's universal level.

Table SP10.1 Compressors designed for use with R134a, their system oil charges, rotation and oil types

Compressor	Oil type	System type	Standard oil charge	Rotation
			Amount, cc (fl. oz.)	
SD5H14	SP-20	TXV	210 ± 15	Either way
		FOT	No standard	
SD7B10	SP-10	TXV	No standard	CW only
SD7H10	SP-20	TXV	No standard	CW only
SD7H13	SP-20	TXV	135 ± 15	CW only
SD7H15/HD	SP-20	TXV	135 ± 15	CW only
		FOT	240 ± 15	
		CCOT	240 ± 15	
SD7H15/SHD	SP-20	TXV	135	CW only
		FOT	240 ± 15	

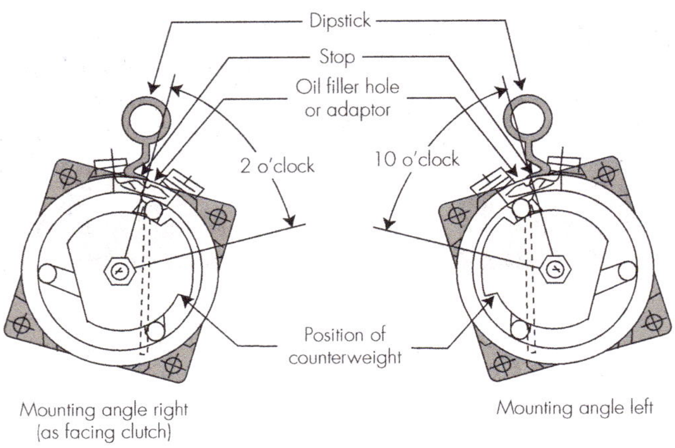

Figure SP10.1 Determining the mounting angle of the compressor

4 Remove the oil filler plug. Then using a socket wrench on the armature retaining nut, turn the shaft clockwise until the counterweight is positioned as shown in Figure SP10.1.

5 Insert the oil dipstick to the stop, as shown in Figure SP10.2, with the angle pointing in the correct direction.

6 Remove the dipstick and count the number of notches covered by oil.

7 Add or subtract oil to meet the specifications shown in Tables SP10.2

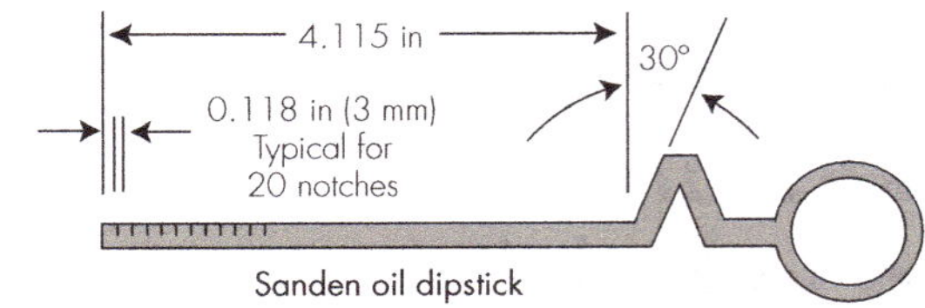

Figure SP10.2 Insertion of the oil dipstick

8 Re-install the oil plug. The seat and new O-ring must be clean and not oiled. Set the torque to 11–18 ft/lb (15–25 Nm, 150–250 kgf/cm).

Table 10.2 Acceptable oil level in increments for R134a compressors using the specified dipstick

Mounting angle (degrees)	Acceptable oil level in increments	
	SD5H14	SD7H15
0	3–5	5–7
10	4–6	6–8
20	5–7	7–9
30	6–8	8–10
40	7–9	9–11
50	8–10	10–12
60	8–10	11–13
90	8–10	16–18

Clutch inspection

1 Measure voltage at the clutch. Low voltage may be due to poor ground or power connections, or to problems with the vehicle electrical system. A good earth requires a tight fit of the field coil retaining snap ring.
2 Measure current draw when the clutch is engaged (see Figure SP10.3). Normal current should be within 3–4.A at 12V DC (allow for slight tolerances):
   - *Overcurrent* – short circuit within the field coil or in the compressor circuit.
   - *No current* – open circuit. If a short or open circuit is found in the field coil, it must be replaced.
3 Measure clutch coil resistance with the circuit not engaged and with no voltage supply, Normal clutch coil resistance should be 3–4 ohms (allow for slight tolerances – see Figure SP10.4).

Compressor repair

Oil check

1 Drain oil from the suction and discharge ports into a suitable container while turning the shaft (clockwise only) with a socket wrench on the armature retaining nut.
2 Measure and record the amount of oil drained from the compressor (see Figure SP10.5 and SP10.6).

Figure SP10.3 Checking the amp draw of the coil – the meter shows 2.73 amps

Figure SP10.4 Checking the resistance of the coil – the meter shows 3.4 ohms resistance

3 Inspect the oil for signs of contamination, such as discoloration or foreign material.
4 Undertake repairs to the compressor.
5 Add the correct amount of new oil to the compressor (see Figure SP10.7). Always use the correct oil for the compressor.

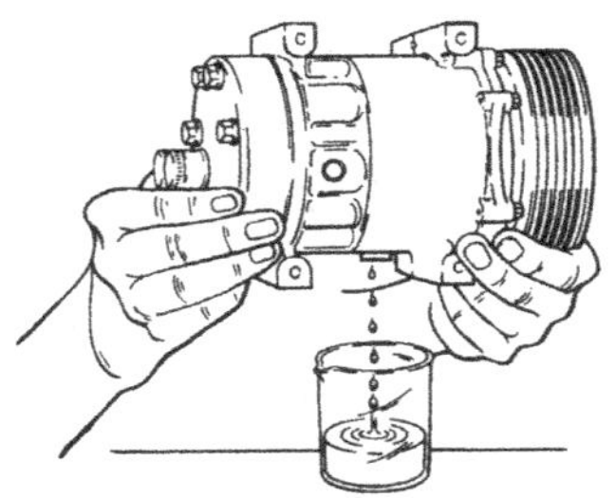

Figure SP10.5 Draining oil from the compressor
Source: Sanden International (Australia) Pty Ltd

6 Re-install the oil plug. The seal and O-ring must be clean and undamaged. Torque should be 11–18 ft/lb (15–25 Nm, 150–250 kgf/cm). Do not cross-thread the oil plug.

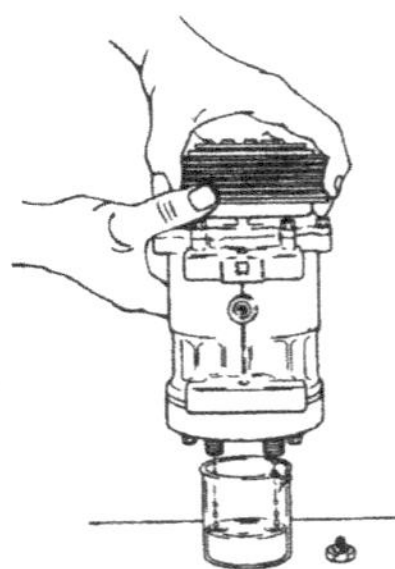

Figure SP10.6 Measuring and recording oil from the compressor
Source: Sanden International (Australia) Pty Ltd

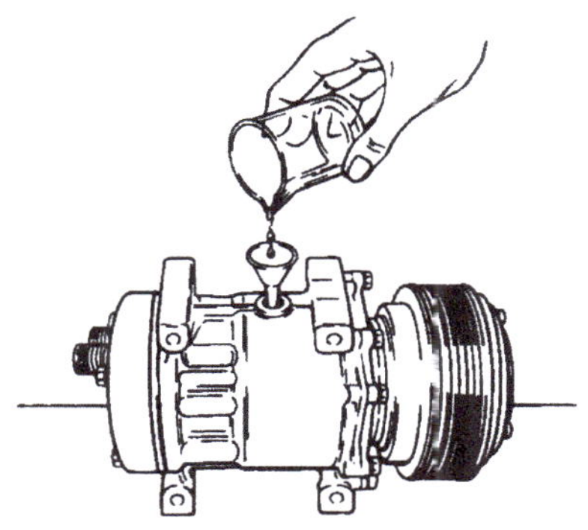

Figure SP10.7 Adding new oil to the compressor
Source: Sanden International (Australia) Pty Ltd

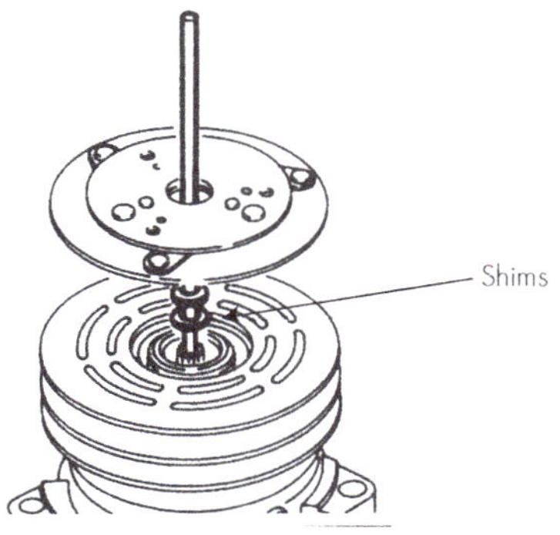

Figure SP10.8 Installing the friction disc
Source: Sanden International (Australia) Pty Ltd

Sanden compressors replaced by a new Sanden compressor of the same type

1 Drain oil from the old compressor; measure and record the amount as described above.
2 Drain oil from the new compressor.
3 Add new oil of the correct type to the new compressor. The quantity should be the same as was removed from the old compressor in step 1 plus oil compensation for replaced components
4 System flushing is not recommended by Sanden. New vehicles come with a full system charge of oil in them.
5 The old system components may need to be cleaned of particles. A component or system flush may be necessary.

Armature assembly removal

1 Insert pins of the armature plate spanner into threaded holes of the armature assembly.
2 Hold the armature assembly stationary while removing the retaining nut with a 19 mm or 14 mm socket.
3 Remove the armature drive plate assembly using the puller. Thread three puller bolts into the threaded holes in the armature assembly. Turn the centre screw clockwise until the armature assembly comes loose.
4 If shims are above the shaft key, remove them now. If they are below the shaft key, the key and bearing dust cover (if present) must be removed before the shims can be removed. See Figure SP10.8
5 Remove the bearing dust cover (if present). Take care not to distort the cover while removing it.
6 Remove the shaft key by tapping it loose with a slotted screwdriver and hammer.
7 Remove the shims. Use a pointed tool and a small screwdriver to prevent them from binding on the shaft.

Rotor assembly removal

1 Remove the internal snap ring for the bearing with the internal snap ring pliers,

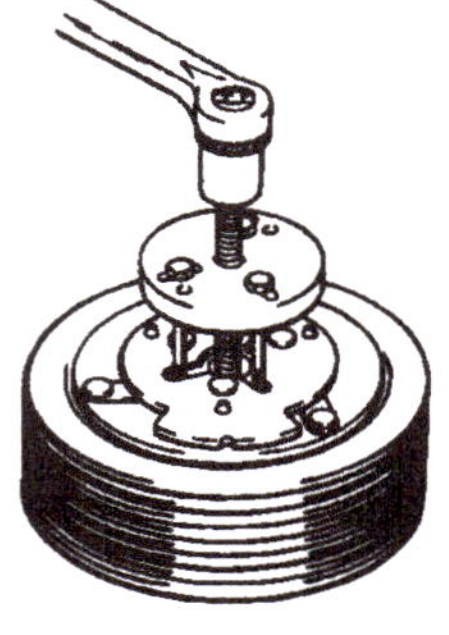
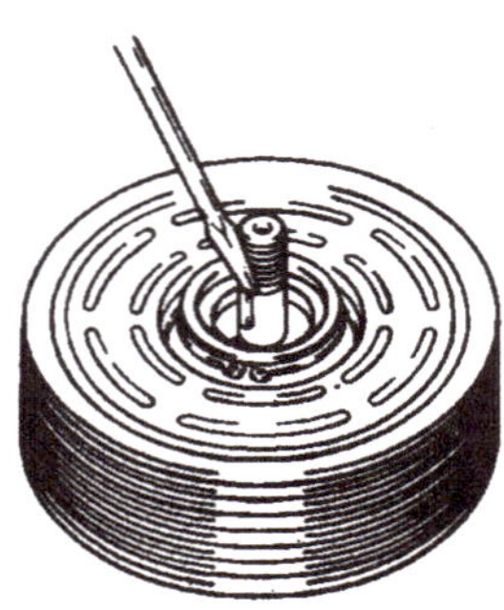

Figure SP10.9 Stages in armature drive plate assembly removal
Source: Sanden International (Australia) Pty Ltd

if it is visible above the bearing. (See Figure SP10.9.)

2 Remove the rotor pulley assembly (see Figure SP10.10):
   - Insert the lips of the jaws into the bearing snap ring groove.
   - Place the rotor puller set over the hub.

3 Align the puller bolt with the compressor shaft.

4 Remove the field coil assembly.

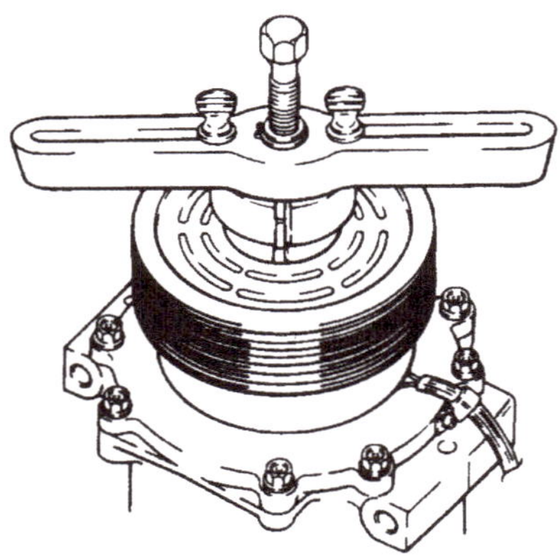

Figure SP10.10 Removal of the rotor assembly
Source: Sanden International (Australia) Pty Ltd

Field coil assembly removal

1 Loosen the lead wire clamp screw with a #2 Phillips screwdriver until the wire(s) can be slipped out from under the clamp.

2 Undo any wire connections on the compressor that would prevent removal of the field coil assembly (see Figure SP10.11).

3 Remove the snap ring or screws and remove the field coil.

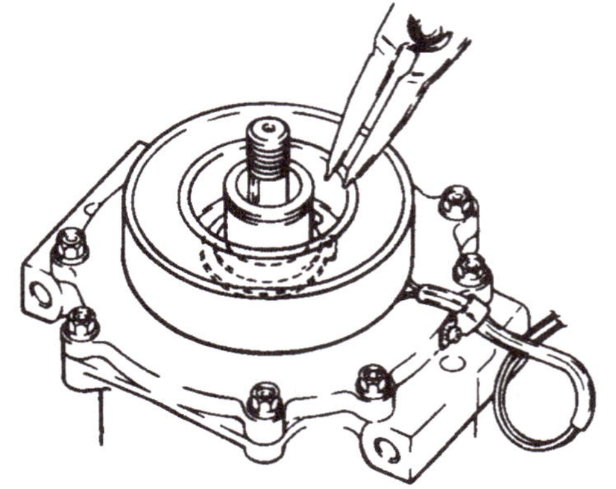

Figure SP10.11 Removal of the field coil assembly
Source: Sanden International (Australia) Pty Ltd

Field coil assembly installation

Reverse the steps of the above section. Note that the protrusion on the underside of the coil ring must match the hole in the front housing to prevent movement. Take care that you correctly locate the lead wire(s).

Armature assembly installation

1 Install the clutch shims. Note that the clutch air gap is determined by shim thickness. The air gap should be between 0.4 and 0.8 mm.
2 Install the shaft key with pliers.
3 Align the keyway in the armature assembly to the shaft key. Using a driver and a hammer or arbor press, drive the armature assembly down over the shaft until it bottoms on the shims. Note the distinct sound change if driving with a hammer.
4 Replace the retaining nut and torque to specifications.
5 Recheck the air gap with a feeler gauge. The gap must be checked in three equal positions around the plate. If the overall gap is outside the specifications, remove the armature assembly and change shims as necessary.

Replacement of the shaft seal

Special tools are required to perform this operation

NOTE

+ The shaft lip seal assembly and felt ring must never be reused. Always replace these components.

1 Ensure that all gas pressure inside the compressor has been relieved.
2 Remove the armature drive plate, armature assembly, rotor-bearing, shaft key and clutch shims.
3 Insert the points of a pair of snap ring pliers into the two holes of the felt ring retainer and pry out the retainer and felt ring.
4 Remove the seal snap ring with internal snap ring pliers.
5 Use the lip seal removal and installation (see Figure SP10.13) tool to remove the lip seal

Figure SP10.12 Installing the armature assembly
Source: Sanden International (Australia) Pty Ltd

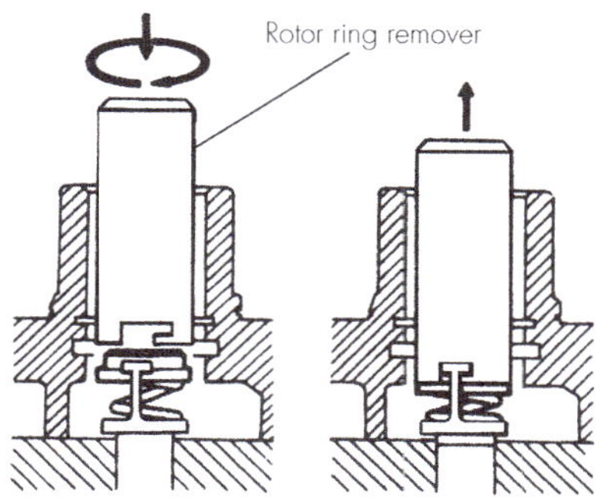

Figure SP10.13 Removing the seal rotor assembly
Source: Sanden International (Australia) Pty Ltd

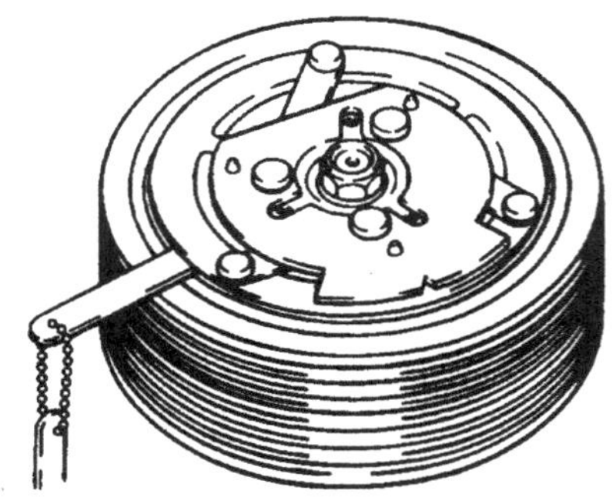

Figure SP10.14 Checking the air gap with a feeler gauge
Source: Sanden International (Australia) Pty Ltd

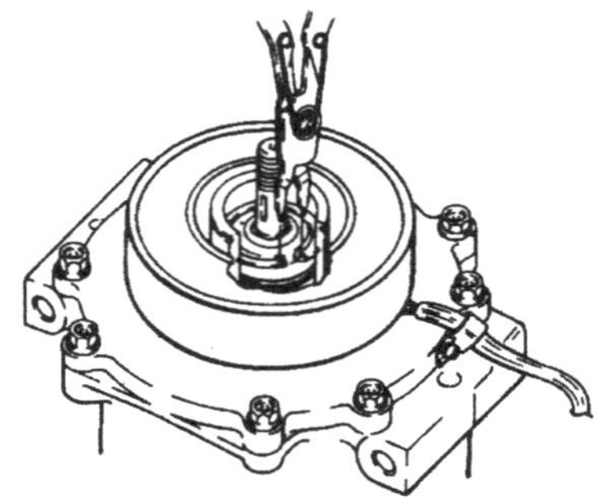

Figure SP10.16 Removing the seal snap ring with internal snap ring pliers
Source: Sanden International (Australia) Pty Ltd

assembly. Twist the tool until the two lips on the tool engage the slots in the lip seal housing, then pull the seal out with a twisting motion.

6 Thoroughly clean out the shaft seal cavity. Debris can be removed using a non-petroleum-based solvent and a lint-free cloth. The area should then be blown out with clean, dry compressed air. Make sure that all foreign material is completely removed.

7 Place the shaft seal protective sleeve over the compressor shaft. Inspect the sleeve to ensure that it has no scratches and is smooth so that the lip seal will not be damaged. Make sure that there is no gap between the end of the sleeve and the seal surface of the shaft.

8 Engage the lips of the seal removal and installation tool with the slots in the new lip seal housing. Make sure the lip seal assembly, and especially the O-ring, is clean. Dip the entire lip seal assembly on the tool into clean refrigerant oil. Ensure that the seal assembly is completely covered with oil.

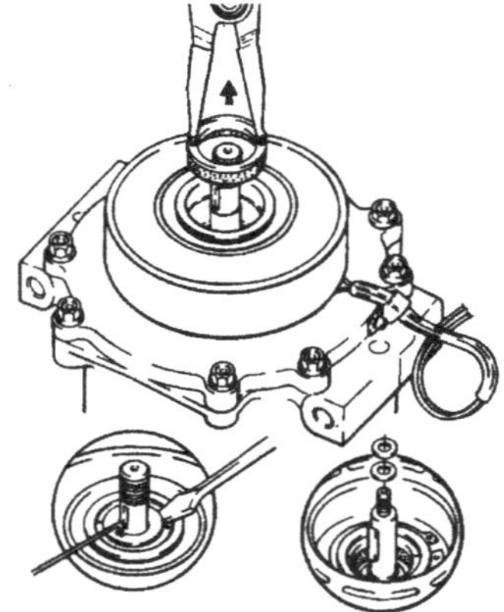

Figure SP10.15 Inserting snap ring pliers into the two holes of the felt ring retainer
Source: Sanden International (Australia) Pty Ltd

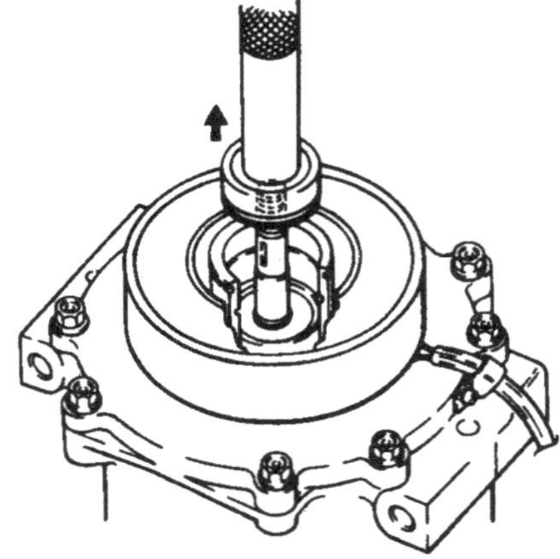

Figure SP10.17 Pulling out the lip seal assembly
Source: Sanden International (Australia) Pty Ltd

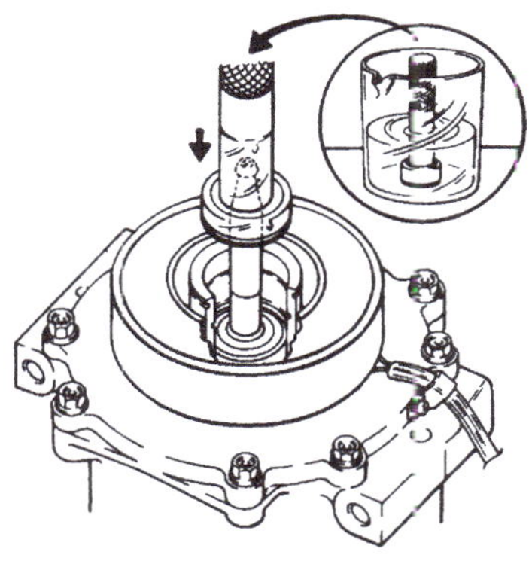

Figure SP10.18 Placing the shaft seal protective sleeve over the compressor shaft
Source: Sanden International (Australia) Pty Ltd

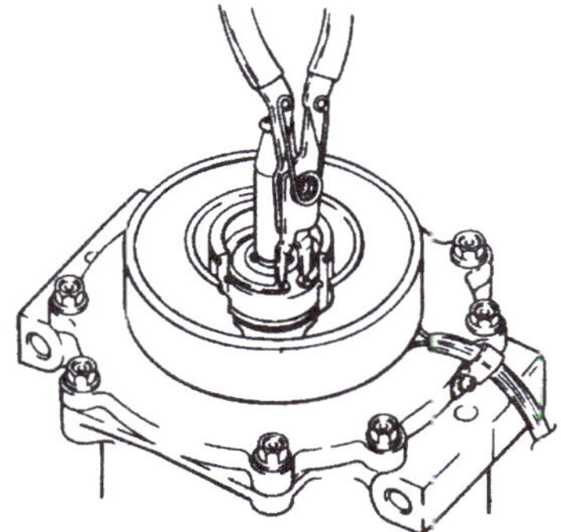

Figure SP10.19 Re-installing the shaft seal snap ring with internal snap ring pliers
Source: Sanden International (Australia) Pty Ltd

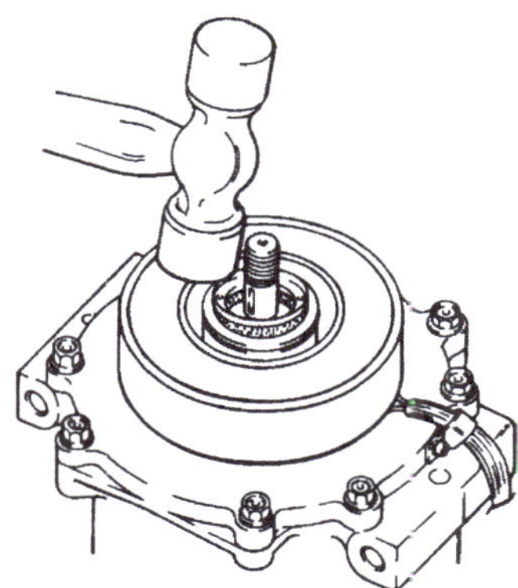

Figure SP10.20 Tapping the new felt ring assembly into place
Source: Sanden International (Australia) Pty Ltd

9 Install the lip seal over the shaft and press firmly to seat. Twist the tool in the opposite direction to disengage it from the seal and withdraw the tool.
10 Re-install the shaft seal snap ring with internal snap ring pliers. The bevelled side should face up (away from the compressor body). Ensure that the snap ring is completely seated in the groove. It may be necessary to tap the snap ring lightly to seat it in the groove.
11 Tap the new felt ring assembly into place.
12 Re-install the clutch shims, shaft key, rotor bearing dust cover (if used) and armature assembly – as previously described.
13 Check and adjust the air gap as necessary.
14 Re-install the armature dust cover (if used).

Cylinder head removal

1 Ensure that all internal compressor pressure has been relieved.
2 Inspect the cylinder head for fitting or thread damage. Replace it if it is damaged.
3 Remove the cylinder head bolts.

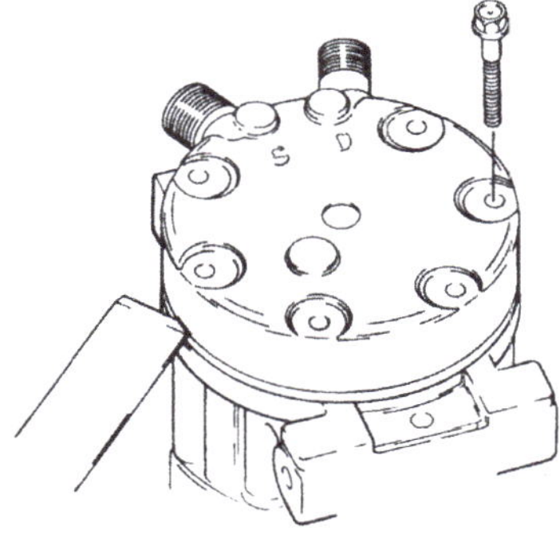

Figure SP10.21 Using a small hammer and gasket scraper to separate the cylinder head from the valve plate
Source: Sanden International (Australia) Pty Ltd

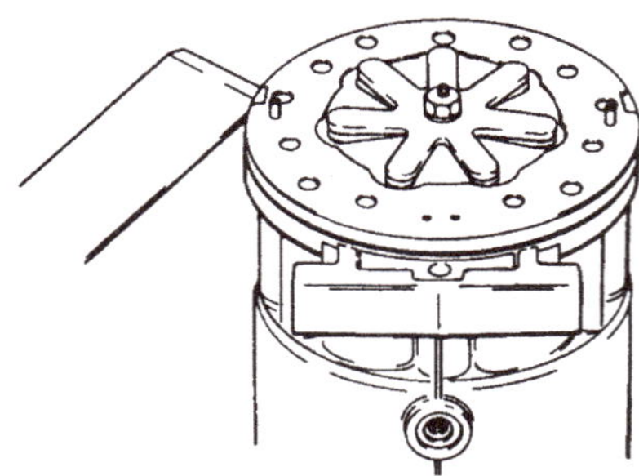

Figure SP10.22 Using a gasket scraper to remove the old head gasket

Source: Sanden International (Australia) Pty Ltd

4 Use a small hammer and gasket scraper to separate the cylinder head from the valve plate. Be careful not to scratch the gasket surface of the cylinder head.

5 Carefully lift the cylinder head from the valve plate.

6 It is recommended that both the head gasket (between the cylinder head and the valve plate) and the block gasket (between the valve plate and cylinder block) be replaced every time the cylinder head is removed. However, if no service is required to the valve plate, it may be left in place. If the valve plate comes loose from the cylinder block, the block gasket must be replaced.

7 Carefully remove the old head gasket from the top of the valve plate with a gasket scraper. Be careful not to disturb the valve plate/cylinder block joint if the valve plate is to be left in place. If the valve plate comes loose from the cylinder block, proceed to valve plate removal, and replace the block gasket.

Valve plate removal

1 Using a small hammer and gasket scraper, carefully separate the valve plate from the cylinder block. Be careful not to damage the sealing surface of the cylinder block.

2 Inspect the reed valves and retainer. Replace the valve plate assembly if any part is damaged.

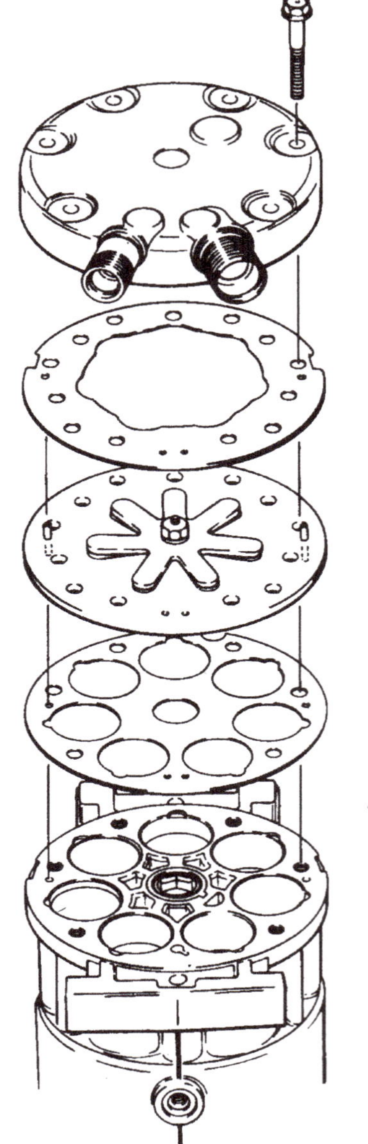

Figure SP10.23 Valve plate and cylinder head installation

Source: Sanden International (Australia) Pty Ltd

3 Carefully remove any gasket material that might be remaining on the valve plate, cylinder block or cylinder head. Do not damage the sealing surfaces of components.

Valve plate and cylinder head installation

Large gasket: the outside diameter of the block gasket is (120 mm) and the sealing face of the block does not have a (115 mm) diameter step.
Small gasket: the outer diameter of the gasket is (115 mm) and the sealing face of the cylinder block has a (115 mm) diameter step.

1 Place the valve plate on the cylinder block with the discharge valve, retainer and nut facing up (away from cylinder block), and location pins properly located in holes.
2 Use a vacuum pump and small tube to remove residual oil from each bolt hole. If this step is not performed, hydraulic pressure can be created when the cylinder head bolts are tightened. This pressure can break the cylinder block.
3 Coat the head gasket with clean refrigerant oil.
4 Install the head gasket over location pins, checking for correct orientation.
5 Install the cylinder head.
6 Install the cylinder head bolts and tighten in a star pattern. Torque first to approximately 20 Nm, then finish by torquing to 35 Nm.
7 Coat the new block gasket with clean refrigerant oil (making sure that it is the oil specified for that particular compressor and system).

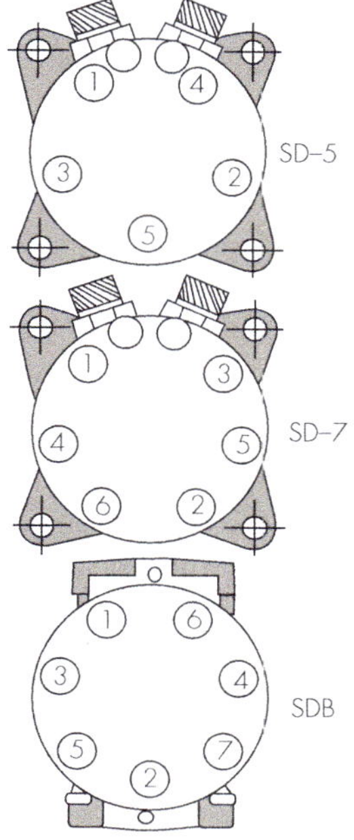

Figure SP10.24 Tighten cylinder head bolts in a star fashion
Source: Sanden International (Australia) Pty Ltd

8 Install the block gasket. Align the new gasket to the location pin holes and orifice(s). The notch (if present) should face the same direction as the oil plug or adaptor.

Thermal protector switch (TPS) testing

Some models of SD compressors are equipped with a bi-metal thermal protector switch (TPS) to protect against abnormally high temperatures.

1 Check the continuity at room temperature. If the switch is open at room temperature, remove and replace it.
2 Check the actuating temperature. Remove the TPS and place it in a container of PAG oil. Heat the oil using an electric hot

plate while monitoring oil temperature. The thermal protector switch should open at 110 to 120°C. Replace it if it does not function properly.

TPS replacement

1. Disconnect all electrical connections.
2. Remove the TPS retaining clip bolt.
3. Spray around the TPS with commercial non-petroleum-based solvent (a volatile type such as 1, 1, 1-trichloroethane or an approved substitute) to loosen the silicone.
4. Remove the TPS with pliers. Use care to avoid deforming the TPS housing, as this can change the temperature setting.
5. Clean silicone residue out of the TPS well with a flat-bladed screwdriver. Wipe out the TPS well with a cloth. Make sure the well area is clean and dry.
6. Apply a dot of silicone RTV, approximately (6 mm in diameter and 3 mm high at the bottom of the TPS well.
7. Install a new TPS, making sure that the lead wires are oriented to the clearance notch. The re-use of a TPS that has been removed is not recommended.
8. Install the TPS retaining clip and bolt. Hold the clip tight against the stop while torquing the bolt to 8–13 Nm.
9. Reconnect all electrical connections and check function.

High-pressure relief valve (HPRV) replacement

Some models of Sanden compressors are fitted with a high-pressure relief valve (HPRV) to protect against damage from abnormally high discharge pressures; this is currently achieved by the use of high- and low-pressure switches or pressure transducers.

When replacing a failed HPRV with a new one, be sure to check whether the air-conditioning system is designed for R12 and is now running R134a. The HPRV and the small O-ring at the threaded portion are different for R134a.

1. Be sure that all gas pressure has been released from inside the compressor.
2. Remove the HPRV.
3. Coat the O-ring of the new HPRV with clean refrigerant oil. The seat and O-ring must be clean and not damaged.
4. Install the new HPRV and torque to 8–12 Nm.

3.11 SERVICE PROCEDURE 11

Servicing the Sanden TR70/90 compressor

Inspection and repair procedures for TR compressors

General

- All repair and service operations should be performed on a clean bench and with the use of clean tools.
- Use genuine parts and correct tools for all repair service work.
- Tighten every nut/bolt to specified torques.
- Clutch pulley and shaft seal repair procedures for TRF/TRS090 and TRF/TRS150 are different from those for TR70/90. Other parts' repair procedures are basically similar.

Oil check and charge procedures

Whenever a system component has been repaired or replaced, or there has been an obvious oil leak, follow the procedures below.

After following recommended refrigerant recovery procedures, remove the compressor from the vehicle, and place it on a clean bench.

Drain oil from the compressor via the suction and discharge ports for as long as possible. Ensure that the maximum amount of oil is drained from the compressor. Rotate the compressor crankshaft several times while draining the oil. The amount drained from the compressor only should be approximately 75–85 mL. This is a system full charge. The rest of the oil is contained in other system components.

Using a total of 130 mL of clean refrigerant oil, charge 65 mL of oil through each of the discharge and suction ports.

Note that there will generally be a certain amount of oil adhering to the internal working surfaces of the compressor. At ambient temperatures of between 20 and 25°C, this amount would be approximately 20 mL. By adding 130 mL of oil the total amount in the compressor will be 150 mL, being the total system charge. This is the factory-charged oil amount of a typical model. As the compressor now has the factory charge of oil, it is necessary to flush oil from the remaining components before fitting the compressor back to the system. If this is not done, there is the possibility of an overcharge of oil, which could result in a performance reduction in the air-conditioning system.

It is obvious from the above Sanden-approved procedure that it would be easier and more cost-efficient if the system were to be flushed to remove the system oil. There are three reasons for this:

- This would give a complete system oil change, with no contamination between new and old oils.
- The system components would be returned to an as-new condition.
- There is no need to estimate the amount of oil in each component as all the old system oil charge is removed.

TR70/90 repair procedures

Clutch and shaft seal service-clutch disassembly

1 Remove the armature-to-shaft fixing bolt (Figure SP11.1). Insert the three prongs of the armature securing tool into the three holes on the armature. Then, holding the tool by hand, remove the clutch shaft bolt with an 8 mm socket.

2 Remove the clutch armature assembly (Figure SP11.2). If the armature cannot be removed by hand, use the puller as follows:

   » Align the armature puller, centre the bolt with the compressor shaft. Finger-tighten the three puller bolts into the threaded holes. Turn the centre bolt clockwise until the armature disengages from the shaft.

Figure SP11.1 Removing the armature fixing bolt

3 Using snap ring pliers, remove the snap ring that secures the clutch rotor to the front boss.

NOTE

+ Take care that the snap ring does not fly out at the point of release from the groove.

4 Remove the rotor pulley assembly (Figure SP11.3). The clutch rotor is a

Figure SP11.2 Removing the clutch armature assembly

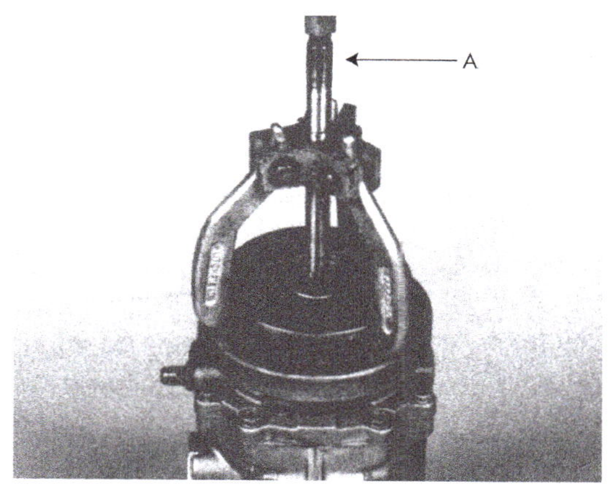

Figure SP11.3 Removing the rotor pulley assembly

tight fit to the front boss. Remove by using a suitable puller as shown in Figure SP11.4.

NOTE

+ Turn handle A (Figure SP11.3) *clockwise.*

5 Remove the clutch coil assembly:
» Loosen the lead wire retaining clamps by using a Phillips head screwdriver, and remove the lead wire from the compressor front-end plate. Disconnect the lead wire from the thermal protector switch.
» Remove the snap ring that secures the clutch coil assembly to the front boss using snap ring pliers.

NOTE

+ Take care that the snap ring does not fly out from its groove.

» Remove the air gap adjusting shims from the top of the compressor shaft.

Figure SP11.4 Removing the front boss

Shaft seal disassembly

1 Remove the front boss using a 10 mm socket, remove the four bolts that secure the front boss of the front end plate as illustrated in Figure SP11.5. Lift the front nose off the compressor shaft, then remove the square-cut O-ring from the front end plate.

Figure SP11.5 Removing the shaft seal

Figure SP11.6 Removing the seal plate

2 Remove the shaft seal by inserting the seal removal and installer tool against the seal assembly. Gently press down against the seal spring and twist the tool until it is engaged into the slots of the seal cage. Lift out the assembly seal.

3 Remove the seal plate (Figure SP11.6). Depress the arms of the seal plate removal and installer tool and install the tool into the inside of the seal plate. Allow the ends of the tool to expand and engage into recesses on the base of the seal plate. Using a pulling action, lift the seal plate out of the front boss.

4 Remove the felt ring from the bore of the front boss using the hook tool.

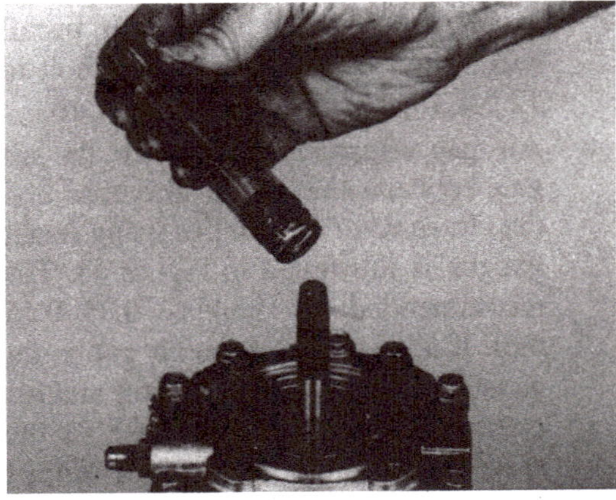

Figure SP11.7 Replacing the seal over the sleeve protector

Shaft seal reassembly

1 Reassemble the complete shaft seal assembly and front boss:
   » Replace the felt ring by inserting a new felt ring into the centre bore of the front boss so that it is correctly in line with the inner wall.
   » Replace the seal plate. With the seal plate installed on the end of the installing tool, apply clean refrigeration oil to the seal plate O-ring. Install the seal plate into the bore of the front boss.

NOTE

\+ Make sure that the seal surface faces upwards.

2 Replace the shaft seal:
   » Clean all tools before replacing the shaft seal.
   » Insert the seal sleeve protector over the compressor shaft.
   » Apply a little oil on the seal sleeve protector.
   » Do not touch the new shaft seal lapping surfaces. Dip the mating surfaces into clean refrigeration oil before processing.
   » Engage the slots of the seal remover and installer tool to the new shaft seal cage. Install the shaft seal by pushing it gently over the shaft seal protector. After pressing it firmly into place, apply further gentle pressure and twist the tool to disengage it from the shaft seal cage, then remove the tool.
   » Apply oil on the carbon face of the shaft seal.
   » Remove the seal sleeve protector.

NOTE

+ Take care that the seal face is not touched, as any form of contamination will result in seal failure. Never reuse any older parts of the shaft kit.

3 Replace the square-cut O-ring. Place a new one in the ring groove of the front-end plate by hand.

4 Install the front boss. Align the recessed part of the front boss or small hole on the front boss as illustrated in Figure SP11.8. Install the front boss carefully on to the shaft, ensuring that the square-cut O-ring has not moved out of place. Then gently press the front boss against the shaft seal spring, pressing on to the front-end plate and into the correct position. Holding the front boss in position, insert the four retaining bolts and tighten by hand. Now apply a 10 mm socket and torque wrench to fully tighten.

Figure SP11.8 Installing the front boss, showing the alignment

NOTE

+ Tighten the bolts in a diagonally progressive order to a torque setting of 15 Nm.

Clutch reassembly

1 Install the clutch coil assembly:
   » Position the back of the clutch coil over the front boss so that the locating nipple on the back of the coil lines up with the locating indentation on the front boss (Figure SP11.9). This ensures a correct angular position for the clutch coil and lead wire.
   » Secure the clutch coil in position with a snap ring, using snap ring pliers.
   » Secure the lead wire with a lead wire retaining clamp using a Phillips head screwdriver.
   » Connect the lead wire with the thermal protector switch.

2 Install the rotor pulley assembly: Place the rotor installer on to the clutch bearing. Press fit the rotor pulley assembly by using a hand press.

Figure SP11.9 Installing the clutch coil on to indentation

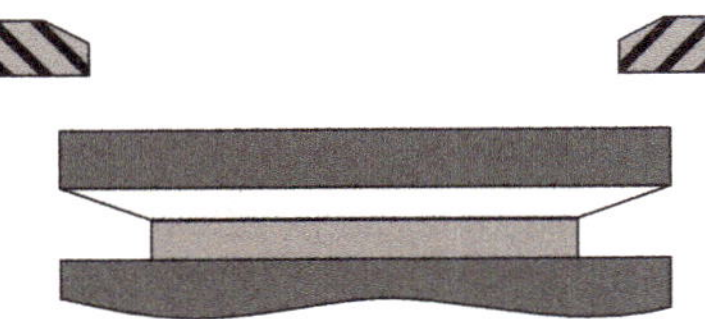

Figure SP11.10 The chamfered edge of the snap ring faces upwards

NOTE

+ The press fit force should not be more than 1.2 tonnes.

3 Install the snap ring into the snap ring groove of the front boss using snap ring pliers.

NOTE

+ Install it so that the chamfered edge of the snap ring faces upwards, as shown in Figure SP11.10.

4 Install the shims: fit the shims on to the shaft. In any case, where more than one shim is used, there is no special order of installing them.

5 Install the clutch armature assembly, aligning the flat on the end of the compressor shaft with the corresponding flat on the armature. Using the armature securing tool and an 8 mm socket, install and torque the armature fixing bolt to 19.5 + 1 Nm (nut type 18 ± 1 Nm).

Figure SP11.11 Air gap adjustment

6 Air gap adjustment: measure the air gap between the clutch armature and the clutch rotor assemblies, using feeler gauges as illustrated in Figure SP11.11. The recommended air gap is 0.35–0.65 mm. If adjustment of the air gap is indicated, remove the fixing bolt and armature clutch assembly (refer to the preceding instructions), then adjust the shim thickness to achieve the recommended air gap.

Thermal protector switch service

Refer to Chapter 10 on TPS testing.

3.12 SERVICE PROCEDURE 12

Servicing the TRF/TRS090 and the TRF/TRS105 Sanden compressors

Inspection procedures

The inspection procedures are the same as for the TR70/90 range. The field replacement parts are shown in Figure SP12.1.

Clutch disassembly

Removal of armature to shaft fixing nut

Procedure is the same as described in Service Procedure 11.

Removal of clutch armature assembly

If the armature cannot be removed by hand, use the puller. Align the puller centre bolt to the compressor shaft. Install (hang) the three puller hooks into the holes on the front plate. Turn the centre bolt clockwise until the armature disengages from the shaft.

Removal of the snap ring

Remove the snap ring that secures the clutch rotor to the front housing.

NOTE

+ Take care that the snap ring does not fly out at the point of release from its groove.

Item	Description
1	Front plate and rotor assembly
2	Clutch field coil
3	Accessory kit
4	Shaft bearing
5	Shaft seal kit
6	Snap ring for shaft seal
7	Thermal protector switch assembly
8	Thermal protector switch fixing plate with lead wire clamp
9	Thermal protector switch fixing plate mounting bolt

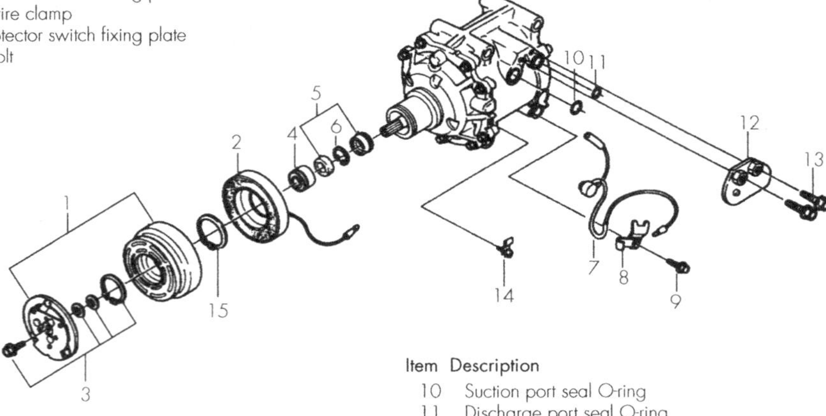

Item	Description
10	Suction port seal O-ring
11	Discharge port seal O-ring
12	Port protective plate assembly
13	Port bolt
14	Clamp with screw
15	Snap ring for clutch field coil

Figure SP12.1 Field replacement parts

Figure SP12.2 Removal of the armature fixing nut

Figure SP12.3 Removing the clutch armature assembly

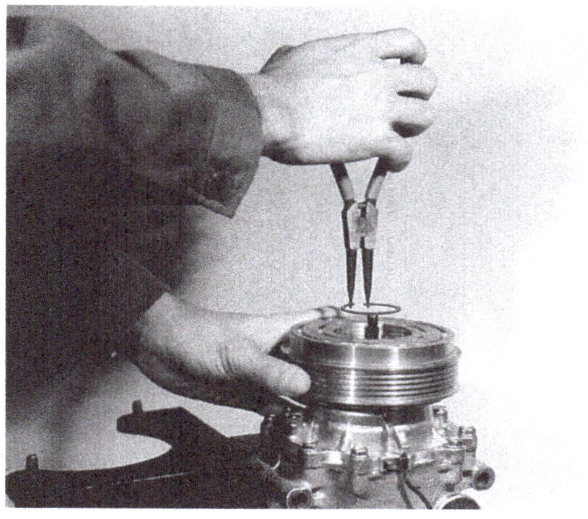

Figure SP12.4 Removal of the snap ring

Removal of the clutch rotor

The clutch rotor fits tightly to the front boss. With the shaft protector in place, remove the rotor with the puller as shown in Figure SP12.5. The clutch bearing is not a serviceable item for the TRF/TRS090 and the TRF/TRS150. The rotor and bearing assembly has to be replaced as one unit.

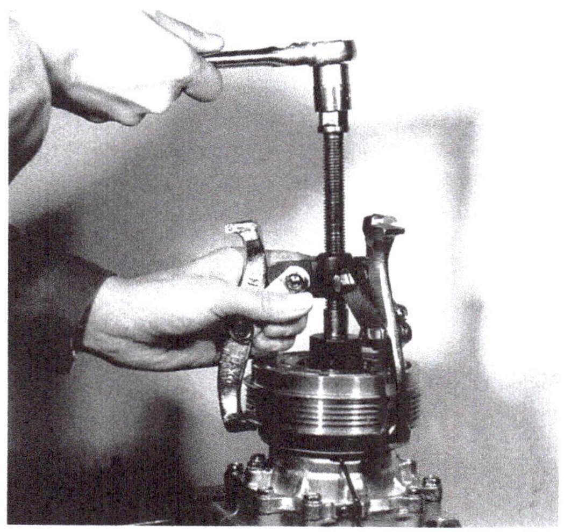

Figure SP12.5 Removal of the clutch rotor

Removal of the clutch coil assembly

› Loosen the lead wire retaining clamp, using a Phillips screwdriver, and remove the lead wire from the compressor. Disconnect the lead wire from the thermal protector switch.
› Remove the snap ring that secures the clutch coil assembly to the front housing using snap ring pliers.

NOTE

+ Take care that the snap ring does not fly out of its groove.

› Remove the air gap adjusting shims from the top of the compressor shaft.

Figure SP12.6 Removal of the clutch coil assembly

Clutch reassembly

Installation of the clutch coil assembly

- Position the back of the clutch coil over the front housing. This ensures the correct angular position of the clutch coil and lead wire.
- Secure the clutch coil in position on the front housing with a snap ring using snap ring pliers.

NOTE

+ Install it so that the chamfered edge of the snap ring faces upwards as shown in Figure SP12.7.

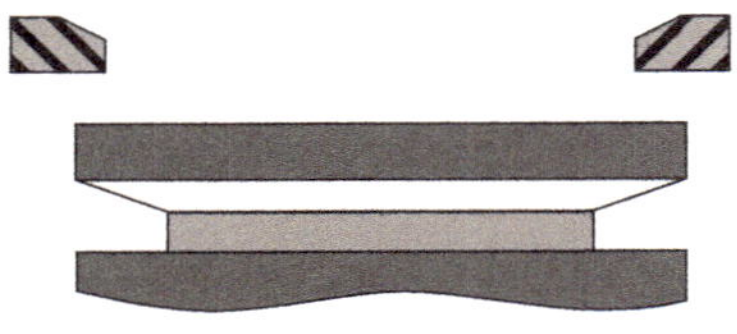

Figure SP12.7 The chamfered edge of the snap ring faces upwards

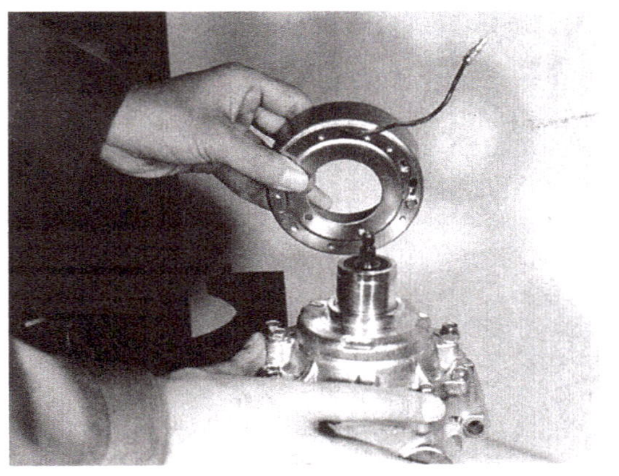

Figure SP12.8 Installation of the clutch coil assembly

Figure SP12.9 Installation of the clutch rotor

- Secure the lead wire with a lead wire-retaining clamp using a Phillips screwdriver.
- Connect the lead wire with the thermal protector switch.

Installation of the clutch rotor

- Place the rotor assembly over the shaft and onto the front housing.
- Place the rotor installer on to the clutch bearing.
- Press-fit the rotor assembly using a hand press.

NOTE

+ Press-fit force must not exceed 400 kg.

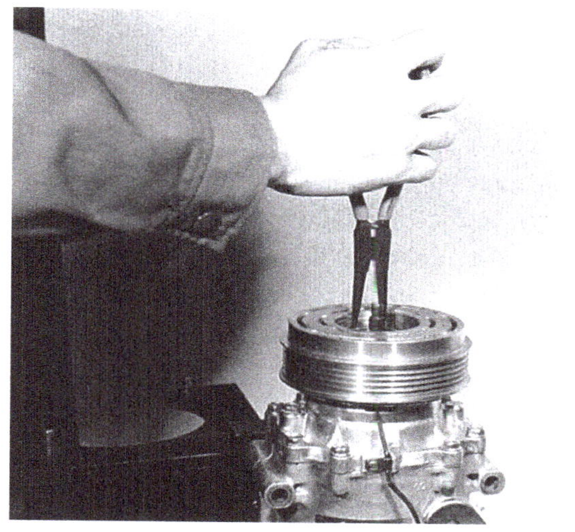

Figure SP12.10 Installation of the snap ring

Installation of the snap ring

Install the snap ring into the snap ring groove of the front housing, using snap ring pliers.

NOTE

+ Install it so that the chamfered edge of the snap ring faces upwards as shown in Figure SP12.7.

Installation of the shims

Refit the shims on to the shaft. In the case where more than one shim is used, there is no special order for installing them.

Installation of the clutch armature assembly

Install the clutch armature, aligning the flat on the end of the compressor shaft with the

Figure SP12.11 Installation of the clutch armature assembly

corresponding flat on the armature. Using the armature securing tool and a 14 mm socket, install and torque the armature fixing bolt to 18 Nm.

Air gap adjustment

Measure the air gap between the clutch armature and clutch rotor using feeler gauges as illustrated in Figure SP12.12. The recommended air gap is 0.35–0.65 mm. If adjustment of the air gap is indicated, remove the fixing nut and armature assembly (refer to the preceding instructions) then add or remove shims as necessary to achieve the recommended air gap.

Figure SP12.12 Air gap adjustment

Shaft seal replacement

The replacement of the shaft seal is not possible without the genuine tools from Sanden. There is a special shaft bearing extraction and seal removal tools. Without these tools the dismantling of the front of the compressor must be untaken. If this is to be done every component must be marked as to its position during replacement. The eight front mounting plate bolts are removed, and the seal is located under the front housing bearing.

NOTE

+ The component parts must be placed back in exactly the same position as then it was dismantled.

Thermal protector switch service

Never re-install a thermal protector switch. Rather, replace it with a new unit if necessary. Refer to Service Procedure 10.

Answers to self-check questions

CHAPTER 1

1 1940.
2 The air is cooled and cleaned, the humidity lowered and the air circulated.
3 Eye protection (safety glasses), overalls with long sleeves, and hand protection.
4 Splash large quantities of cool tap water on the area for several minutes. Cover the area; seek medical advice.
5 Heavier.
6 Keep cool out of direct sunlight and ensure contents are clearly labelled.
7 Système International (the metric system).
8 In grams.
9 170 grams.
10 Quantity (joules) and intensity (degrees) of heat.
11 Minus kPa or mm/hg (mm of mercury).
12 Technicians are required to have the appropriate licence, equipment and to maintain service records.
13 $1100.
14 $5500.
15 Yes (licence CL100).

CHAPTER 2

1 Nitrogen.
2 Between the troposphere and the stratosphere.
3 15–25 km.
4 Dobson units.
5 Three oxygen atoms.
6 A wind that circulates around Antarctica.
7 Antarctica.
8 Increased UV radiation causing damage to plant life, plankton and damage to eyes, skin and the immune system.
9 Chlorine and halons.
10 Carbon dioxide (CO_2).
11 Fire extinguishers.
12 16 years.
13 HFO1234yf.
14 2050.
15 The environment, social issues and economics.

CHAPTER 3

1 37 °C.
2 In Joules.
3 Convection, evaporation and radiation.
4 Hot to cold.
5 Temperature, humidity and air movement.
6 The ambient air is 75 per cent saturated with moisture and can only absorb 25 per cent more.
7 Conduction, convection and radiation.
8 Any heat that can be felt or measured.
9 Latent heat is the quantity of heat required to cause a change of state.
10 Ambient air, sunlight, engine heat, number of passengers, glass area, colour of vehicle etc.
11 Fahrenheit, Celsius, Kelvin and Rankine.
12 Celsius.
13 101.4 kPa
14 Absolute pressure takes into account atmospheric pressure and at rest (at sea level) an absolute gauge would read 101.4 kPa; gauge pressure starts at zero at sea level.
15 92.5 °C.

CHAPTER 4

1 –26°C to –30°C.
2 High-pressure superheated vapour.
3 Tube and fin, serpentine or modine, multi-flow and parallel flow.
4 It stores refrigerant until required, it also filters impurities and removes moisture from the refrigerant.
5 Low-pressure heat-laden vapour.
6 It lowers the pressure of the liquid refrigerant so that it begins to boil and absorb heat from the evaporator.
7 The muffler dampens noises and pulsing from the refrigerant, it is usually found on the inlet side of the compressor.
8 Between the condenser and the TX valve (sometimes in the receiver dryer). It is used to switch the compressor off if system pressure becomes too high.
9 To switch the compressor off if system pressure falls below 35kPa. This prevents the compressor running without lubrication.
10 0°C to 4°C.
11 A surge diode prevents damage to vehicle electronic control units due to spike voltage.
12 Vans, people movers and 4WDs.
13 On the side of the compressor.
14 High and low pressure cut-out and fan operation.
15 On the low-pressure side of the system between the evaporator and the compressor.

CHAPTER 5

1 To prevent cross-contamination of different refrigerant types.
2 R1234yf and CO_2 (R744).
3 The components of the blend may fractionate (separate into its individual components). This may cause the components to leak from the system at different rates due to the different molecule size.
4 11 days.
5 PAG oil.
6 Yes, the pressures are approximately 100kPa lower for R1234yf.
7 Hygroscopic, costly and may cause irritation to skin.
8 TX valves will require changing.
9 Either CO_2 or dry chemical fire extinguishers.
10 To lubricate the compressor and compressor seals.

CHAPTER 6

1 Pressure and temperature.
2 High-pressure reading and equivalent temperature and condenser outlet temperature.
3 TX valve, condenser, service valves, high-pressure switch and receiver dryer/accumulator.
4 Nylon barrier.
5 Any O-rings that are disturbed.
6 The compressor should be removed, drained, flushed and the oil replaced with the appropriate type of oil.
7 Fit auxiliary fan/fans, modify it/them to operate when the compressor is operating.
8 Tag/label/decal.
9 300 grams.
10 50 grams.

CHAPTER 7

1 Less than 10 ppm (parts per million).
2 Absorbs moisture.
3 Receiver dryer or accumulator.
4 Lowering the boiling point of the moisture by placing the system under a negative pressure (vacuum).
5 Installing the receiver dryer last, capping all new parts, never working in wet or humid areas, keeping refrigerant oil in sealed containers and evacuating systems for a minimum of 45 minutes.
6 Moisture can block TX valves and orifice tubes, and can react with the refrigerant to form acids.
7 To compensate for the oil left in the replaced component.
8 A minimum of 45 minutes.
9 When the system is being retrofitted or shows signs of excess moisture contamination.
10 15.6°C.
11 To store, dry and filter refrigerant.
12 To allow oil to return to the compressor if the filter becomes blocked.
13 Flare fitting, barb fitting and O-rings.
14 Silica gel and molecular sieve.
15 Leaks, blockages, reverse fitment and age.

CHAPTER 8

1 3500 kPa.
2 The low-pressure gauge can read vacuum or negative pressure.
3 Hand valves should be kept closed.
4 To allow for changes in atmospheric conditions and change in spring tension.
5 To diagnose system problems and to remove and replace refrigerants.
6 Anywhere between the evaporator and the compressor.
7 Between the compressor and the TX valve.
8 A gauge that reads both pressure and vacuum.
9 Remove the glass or plastic cover and using a small screwdriver, adjust the calibration screw until the gauge reads zero with no pressure in the hose.
10 The blue is connected to the low-pressure gauge, the red is connected to the high-pressure gauge and the yellow hose is used to connect either refrigerant or vacuum pumps to either the blue or red hoses via the manifold.

CHAPTER 9

1 Lack of cooling.
2 Electronic, UV dye and nitrogen gas.
3 The sensing tip.
4 Most evaporators are under the dashboard and often inaccessible.
5 Below the joint or connection as refrigerant is heavier than air.
6 Inserting the probe of the electronic leak detector in a drain tube or by looking for stains in a drain tube if you are using UV dye.

7 It can be added at service so any future leaks can be detected readily.
8 100 grams.
9 Add nitrogen until the pressure reads between 1200–1400 kPa.
10 The system should be rechecked for further leaks.

CHAPTER 10

1 Create a low-pressure area at the inlet and a high-pressure area at the outlet.
2 Reciprocating piston, scroll and vanerotary.
3 By an electromagnetic clutch.
4 Lack of system performance and fluctuating gauge needles.
5 One.
6 By changing the angle of the swash plate.
7 In most cases it has no need to cycle a magnetic clutch.
8 V-belt, cogged v-belt and multi-ribbed serpentine belts.
9 390–440 Nm.
10 Higher than normal low-pressure gauge readings and lower than normal high-pressure gauge readings.

CHAPTER 11

1 TX valve.
2 Internally and externally equalised.
3 High-pressure liquid.
4 The ball and pad.
5 0–2°C.
6 Low-pressure.
7 To throttle and modulate refrigerant flow.
8 Fixed orifice tube (FOT).
9 The compressor cycles for longer or shorter durations.
10 To ensure the system lubrication is circulated. This prevents the TX valve or FOT from sticking and keeps seals moist to prevent leaks.

CHAPTER 12

1 Cartridge, ceramic and blade.
2 Current flow stops and the circuit will not operate.
3 In amps.
4 To keep the temperature inside the cabin regulated and to de-ice the evaporator.
5 30–35 kPa.
6 2400–3200 kPa.
7 A high-temperature cut-out switch that prevents compressor operation if it gets too hot.
8 Vacuum reserve tank and check valve.
9 To ensure the oil is circulated through the air-conditioning system to prevent seals drying out and refrigerant leaking.
10 3 to 4 amps.

CHAPTER 13

1 Cross-flow and vertical-flow.
2 To contain the coolant and pressurise the cooling system.
3 To increase the boiling point.
4 It allows coolant to return back into the cooling system from the surge or overflow tank, thus preventing the hoses from collapsing.
5 To allow a quick warm-up of the engine.
6 To prevent thermal shock to the radiator.
7 A drive belt, timing belt or electric motor.
8 Electric, fixed drive (belt) and viscous coupling.
9 Anti-freeze, anti-boil and corrosion inhibitor.
10 Air intake, core section (heater and evaporator) and air distribution.

CHAPTER 14

1 Heating and cooling are pre-set by the driver and controlled by an ECU.
2 Cabin, ambient, engine coolant, evaporator, sun load, blower speed, temperature-setting input and blend door position sensors.
3 Between 46 and 50 per cent.
4 A 20 MΩ ohmmeter.
5 Behind the air intake grill just in front of the condenser.
6 Photo-voltaic.
7 Decrease.
8 To mix hot and cold air in the correct proportion to achieve the temperature input of the driver.
9 The compressor surge diode.
10 Display fault codes, stream live data and display performance graphs.

GLOSSARY

accumulator a low-pressure refrigerant storage and filtering device

air conditioning a process of cooling, cleaning, dehumidifying and circulating air

ambient air temperature the temperature of the air around the refrigerant system

ATC automatic temperature control

atmospheric pressure the pressure of the air that is around us; 101.4 kPa at sea level

auxiliary fans extra fans used to control air flow

bar a metric pressure equalling 100 kPa

belt slip sensor electronic sensor that detects a slipping drive belt

binary switch a dual pressure switch, e.g. low-high or mid-high-pressure switch

blend door used to mix hot and cold air together

blended refrigerant or blend a refrigerant that has more than one component

boiling point the point at which a liquid starts to change to a vapour

CFC chlorofluorocarbon or Refrigerant R12

capillary tube refrigerant-filled tube from TX valve to evaporator outlet; senses temperature

Celsius metric temperature scale with water freezing at 0°C and boiling 100°C

change of state ice changing to liquid or liquid changing to vapour and vice versa

charging hose the service hose used to put refrigerant into the system

clutch a device through which drive or no drive can be obtained

compound gauge low-pressure gauge that indicates vacuum and pressure

compressor a vapour pump that draws in low pressure and pumps high pressure

condensation the process by which a vapour changes to a liquid

condenser component where high-pressure vapour changes to high-pressure liquid

condensing pressure the pressure required to change refrigerant vapour to a liquid

contaminant dirt or moisture that can enter the air-conditioning system when opened

convection heat transfer by liquid or vapour

dehumidify lower the moisture content

density mass per unit volume

desiccant drying agent found in receiver dryers or accumulators

discharge line line carrying vapour from the compressor to the condenser

discharge pressure pressure that is shown on the high-pressure gauge

displacement volume of a cylinder – compressor or vacuum pump

drive pulley rotates the compressor clutch

ECC electronic climate control

electromagnetic clutch an electrically-operated device that enables connection of the compressor to the drive belt

electronic leak detector a refrigerant leak detector – finds leaks as small as 5 grams per year

EMF electromotive force is voltage. When voltage passes through a coil and is turned off it causes a high EMF or voltage spike (up to 40v) and this can damage sensitive components

equalisation pressure is equal on both sides of a unit

ester oil polyoleoester (POE) synthetic oil

evacuate removing air and moisture from the system under a vacuum

evaporation a liquid changing to a vapour

evaporator component where liquid refrigerant changes to a vapour

expansion valve component (TX valve) that restricts refrigerant flow

Fahrenheit an imperial scale for measuring temperature

field coil electromagnetic coupling for the compressor drive

filter removes contaminants from the refrigerant

flooding an over-supply of refrigerant in the evaporator

flush removing all traces of refrigerant and oil out of the system

fractionation the separation of a compound into its various parts, e.g. blends

freezing point the temperature at which a liquid becomes a solid

gauge calibration Setting the pressure gauge to read zero with no pressure on it

gauge pressure a measurement of system pressure that does not take into account atmospheric pressure, i.e. at sea level not connected to a system it should read zero

gauge set two gauges set on a manifold

glide the time taken for a refrigerant to take on or remove heat

greenhouse effect the warming of the atmosphere due to the increase of carbon dioxide

head pressure pressure that is shown on the high-pressure gauge

heat exchanger component that allows heat to be removed or added

heat load amount of heat within an area, e.g. a vehicle interior

heater core a small radiator used to heat the vehicle interior

HFO1234yf also known as R1234yf, is a new refrigerant to be used by most manufacturers after 2011, it is non-ozone depleting and has a very low impact on the greenhouse effect.

high-pressure gauge a gauge that connects to the high-pressure side of the air-conditioning system

high-pressure switch stops the compressor operation when the pressure is too high

high-side gauge see high-pressure gauge

human comfort zone the ambient temperature and humidity at which person feels comfortable

humidity the amount of moisture in the air

hydrocarbon a compound containing hydrogen and carbon

in/Hg an imperial measure of vacuum

Joules a measure of heat energy

kg/cm^2 a metric pressure equalling 100 kPa

kPa a metric pressure

latent heat sometimes called hidden heat, it is the heat that is absorbed while a liquid changes state to a gas without a rise in temperature, it is unable to be measured

leak detector a device for locating refrigerant leaks

liquid line the line between the receiver drier and the TX valve

low-pressure gauge a gauge that connects to the low-pressure side of the air-conditioning system

low-pressure switch stops the compressor when the pressure is too low

magnetic clutch an electromagnetic clutch used to drive the compressor

manifold a device for holding the pressure gauges

manifold hose service hoses for removing and replacing refrigerant

micron a measure of deep vacuum

mm/Hg a metric measure of vacuum

modulation act of opening and closing a restriction-adjusting refrigerant flow

molecular sieve drying agent found in receiver driers and accumulators

OEM original equipment manufacturer

overcharge too much refrigerant in the system

ounce an imperial measurement of weight

ozone earth's protective layer for UV rays

PAG oil polyalkylene glycol oil – a synthetic oil

partial vacuum a state that is less than atmospheric pressure

performance testing checking system operation

pound an imperial measurement of weight

pressure drop pressure lost while moving through a component

pressure transducer an electronic device that measures pressure and sends a varying voltage to the ECU

psi imperial pressure measurement (pounds per square inch) equalling 6.895 kPa

psig imperial pressure shown on the gauge

purge remove all refrigerant from the system

quick-release connectors service fittings used for R134a, blended and 1234yf refrigerants

R12 an ozone-depleting refrigerant

R134a a non-ozone depleting refrigerant

R1234yf a replacement refrigerant for R134a that is a non-greenhouse causing substance

radiation transfer of heat through air or space

ram air air forced through the condenser by vehicle movement

receiver dryer component that removes moisture and contaminates Freon refrigerants

reed valve a valve found in the compressor used as inlet/exhaust valves

refrigerant the compound used in air-conditioning systems to cool the air by removing heat energy from the evaporator to the condenser

refrigerant identifier a machine that identifies the type of refrigerant found in a particular system

relative humidity percentage of moisture found in the air

resistor a device that controls electrical flow and voltage

retrofit change from one refrigerant to another

Roc Oil 68 a polyalphaolefin-based synthetic refrigerant oil

saturated liquid a liquid that is carrying as much heat as possible just before it changes state

Schrader valve an air-conditioning valve similar to a tyre valve used for system access

sensible heat temperature change that is able to be measured

service port used to gain system access

sight glass a small glass window showing refrigerant liquid flow

silica gel a desiccant that is now not used

specific heat the heat required to change temperature by 1 degree

stratosphere the layer of air upper most from the Earth's surface

sub-cooled liquid a liquid that still has the ability to absorb heat. e.g. water between 0 and 100°C

sub-cooling cooling a substance below its saturation temperature

suction line the low-pressure return line

suction pressure the pressure shown on the low-pressure gauge

sun load sensor senses the sun enabling the system to work harder in direct sunlight

superheat added heat to a refrigerant once it has changed state to a vapour

surge diode an electrical spike voltage safety device

thermal protector a safety switch that stops the compressor if casing temperature is too high

thermistor electronic temperature sensor

thermostat senses evaporator temperature and controls compressor operation

thermostatic expansion valve TX valve – a restriction that allows a change of state

thermostatic switch senses evaporator temperature and controls compressor operation (similar to thermistor)

throttling restricting the refrigerant flow

transducer an electronic device that measures pressure and sends a varying voltage to the ECU

troposphere the layer of air closest to the Earth's surface

ultraviolet (UV) dye a method of checking for refrigerant leaks

ultraviolet (UV) radiation a damaging form of radiation from the sun

vacuum a state of absence of pressure usually stated as –101.4kPa

vacuum pump a pump used to remove pressure

vacuum servo unit a vacuum control that opens and closes flaps and doors

vapour a substance in its gaseous state

variable displacement compressor has a movable axial plate that varies the stroke of the compressors pistons and so varies pressure

venturi action this is where a restriction to air flow causes the air speed to increase causing a drop in air pressure in the middle of the restriction (a partial vacuum)

INDEX